The Cato Institute

The Cato Institute is named for the libertarian pamphlets *Cato's Letters,* which were inspired by the Roman Stoic Cato the Younger. Written by John Trenchard and Thomas Gordon, *Cato's Letters* were widely read in the American colonies in the early eighteenth century and played a major role in laying the philosophical foundation for the revolution that followed.

The erosion of civil and economic liberties in the modern world has occurred in concert with a widening array of social problems. These disturbing developments have resulted from a major failure to examine social problems in terms of the fundamental principles of human dignity, economic welfare, and justice.

The Cato Institute aims to broaden public policy debate by sponsoring programs designed to assist both the scholar and the concerned layperson in analyzing questions of political economy.

The programs of the Cato Institute include the sponsorship and publication of basic research in social philosophy and public policy; publication of major journals on the scholarship of liberty and commentary on political affairs; production of debate forums for radio; and organization of an extensive program of symposia, seminars, and conferences.

CATO INSTITUTE
747 Front Street
San Francisco, California 94111

A Property System Approach to the Electromagnetic Spectrum

A Property System Approach to the Electromagnetic Spectrum

A Legal-Economic-Engineering Study

Arthur S. De Vany, Ross D. Eckert, Charles J. Meyers, Donald J. O'Hara, Richard C. Scott

With a Foreword by William H. Meckling

CATO PAPER No. 10

CATO
INSTITUTE
San Francisco, California

Reprinted, with permission, from the *Stanford Law Review,* Volume 21, pages 1499–1561.

ISBN: 0-932790-11-9

Printed in the United States of America.

CATO INSTITUTE
747 Front Street
San Francisco, California 94111

CONTENTS

FIGURES

CONTENTS

A Note from the Authors

This essay is the product of the collaboration of the authors while consultants to the staff of the President's Task Force on Communications Policy. We thought it useful to revise our work and make it available to a larger public. However, the views presented here are those of the authors alone, and do not necessarily coincide with the views of the Task Force, the Executive Office of Telecommunications Management, or the General Electric Company.

Our purpose in publishing this essay is twofold: First, each of us believes that the probability for improved use of the spectrum through market allocation is sufficiently high to warrant congressional approval of an experiment or pilot project testing the property system proposed here. We are, in short, seeking public support for our proposal. Second, apart from the merits of the proposal put forth here, we hope that the essay will be of general interest to lawyers, economists, and engineers concerned with interdisciplinary research. The multifaceted problems to which we address ourselves in this essay could not have been solved by any one of us alone; it was only the collaboration of three disciplines that permitted us to attempt a coherent solution to the legal, economic, and engineering problems arising from the market allocation of spectrum use. Moreover, where two or more possibilities existed for solving certain technical problems (for example, radio interference), each solution had its differing economic consequences and each solution posed differing problems for the legal structure. In short, we found ourselves in a sort of Three Musketeers relationship of one for all and all for one—but, given our subject matter, that is no doubt too romantic. More appropriate, perhaps, are the words of Benjamin Franklin: "We must all hang together, or assuredly we shall all hang separately." Believing this valuable lesson to be demonstrated in our essay, we found independent grounds for making it available to a wider public.

FOREWORD

The invention of radio and the subsequent development of institutions for controlling rights to radiate provide an illuminating case history for scholars interested in the study of government behavior and political processes. Radio's debut exposed a previously unrecognized form of physical incompatibility,[1] namely, the potential that energy emissions have for interfering with one another. Physical incompatibility among individual actions is, of course, an all-pervasive phenomenon, not something peculiar to the emission of energy. What is interesting about the case of emission is not the fact of physical incompatibility itself, but the way the government has chosen to deal with that problem.

Whenever actions that individuals want to take are physically incompatible, some means must be found for resolving the conflict of interest that inevitably follows. Some system must be devised for specifying which individuals are to be able to take what actions. In general, society ends up resolving this conflict by assigning "rights" to take actions to individuals or groups. By their very nature, such rights involve constraints on what actions others can take. The police powers of the state are called upon to ensure (with high probability) that individuals will be able to take the actions to which they have been assigned the rights.

1 In his seminal work on property rights, Ronald Coase uses the word "harm" to describe what I have called physical incompatibility. I prefer the notion of physical incompatibility. "Harm" can be inflicted in two entirely different ways. Harm *can* result from the physical consequences of actions, but "harm" can also result simply from competing offers. ("Harm" is a value concept.) It is important analytically to keep these two kinds of "harm" distinct. Harm due to physical consequences of actions (physical incompatibility of actions) implies imposition of constraints on what acts individuals can take, but "harm" in the sense of reducing values through competing offers does not imply such constraints. Physical incompatibility seems preferable because it highlights the phenomenon that really lies at the heart of the problem.

Rights systems designed to solve the physical incompatibility problem come in a variety of shapes and sizes. One important way they can differ is in the role that government plays. Governments almost without exception undertake the task of enforcement of rights. Governments also generally provide through the courts a means for adjudication of disputes over rights. In the case of rights to radiate, the government has by statute arrogated to itself a much larger role. In particular, a government bureaucracy, the Federal Communications Commission, has been granted the right to decide:

1. What services will be permitted to use various portions of the frequency spectrum;
2. Who will be allowed to offer those services, i.e., who will be permitted to use the spectrum;
3. What kinds of equipment users will be permitted to employ.

What has evolved is a highly centralized frequency spectrum management system, a system that in economic terms can only be described as disastrously inefficient. Much of the spectrum has been relegated to the status of communal property. The class of service permitted in various frequency ranges (mobile, citizens band, etc.) has been specified, but no limit has been placed on the number of users. Nowhere in the spectrum are users forced to consider the costs of exercising emission rights. Under the guise of selling stations, broadcast rights are de facto transferable (subject to FCC approval), but broadcasters do not compete with other potential users for the frequencies that are assigned to broadcasting. In the rest of the frequency spectrum, rights are either not transferable at all or they have zero value because the FCC puts no limit on the number of users. Because users perceive no cost to their use of emission rights, they have no incentive to economize on spectrum utilization. Meanwhile, regulation takes the form of specification of inputs rather than outputs, so that users cannot adopt spectrum-economizing technological innovations without the permission of the FCC, even if they have an incentive to do so. In sum, spectrum management is so bizarre that none of us can even imagine what efficient utilization of the frequency spectrum would look like.

Apparently, however, government's appetite for control in this industry has not been satisfied by its role as commissar for spectrum management. The FCC (with the blessing of Congress) uses its authority to grant emission rights as a club to police TV and radio programming, even though programming has nothing to do with the problem of interference. Without statutory authority, the FCC extended its regulatory arm to cable TV, a communications medium that actually makes no use of the frequency spectrum. Cable TV, of course, is a competitor to broadcast TV, and what the FCC has done is to place constraints on the behavior of a new competitor in order to protect the establishment the FCC itself created. This despite the fact that the substitution of cable for broadcasting could make valuable frequencies now used in broadcasting available for other uses. The FCC for all practical purposes has prohibited the development of broadcast pay TV, while the Congress simultaneously has created and lavishly subsidizes a highly politicized so-called "public" radio and TV industry, which caters to upper-income viewers.

The challenging social science question raised by this whole sordid affair is why political democracy has created and perpetuates a system of rights that is not only grossly inefficient but also presents a real threat to freedom of speech. Even if we accept the view that the system sprang up initially because an economic problem was mistakenly diagnosed as a technical one, how can we explain the almost total lack of interest in reforming it today?

To my knowledge, the suggestion that spectrum rights be made private property was first broached in print by Leo Herzel twenty-eight years ago in the *University of Chicago Law Review.*[2] It has now been twenty years since Coase published his famous article, "The Federal Communications Commission," in the *Journal of Law and Economics.*[3] By 1963 Ronald Coase, Jora Minasian, and I had written and circulated in the FCC, and elsewhere in government, a lengthy report prepared for the National Aeronautics and Space Administration. Our report delineated a system of rights in frequencies that contained all of the essential elements of the system proposed here by De Vany, Eckert, Meyers, O'Hara, and Scott. Moreover, we have witnessed in economics during the last twenty years a flood of articles analyzing the social role of proper-

2 Leo Herzel, "'Public Interest' and the Market in Color Television, Regulation 18," *University of Chicago Law Review* 18 (1951):802–16.

3 Ronald Coase, "The Federal Communications Commission," *Journal of Law and Economics* 2 (Oct. 1959):1–40.

ty rights. Despite all this, it is fair to say we are no nearer to reform of the system of rights in radiation today than we were when Herzel first proposed the idea.

I am personally very pleased that the Cato Institute sees the value of disseminating to a wider audience the results of careful professional research of the sort typified by this article. I must confess, however, that it is hard for me to see at this juncture how we can hope to effect any significant reform in the system of rights to radiate. So many vested interests in the existing system have been created, including particularly the politicians and bureaucrats themselves, that only some dramatic upheaval could evoke the political action required.

October 1979

William H. Meckling
Rochester, New York

INTRODUCTION

Lawyers and economists, as well as members of the business community, have recently shown a growing interest in the regulatory arrangements controlling the rights of individuals and business firms to use the electromagnetic spectrum.[1] These arrangements had their origin in the confused conditions of 1925–27, when rapidly increasing interference threatened to destroy the utility of radio communication. Congress responded by enacting the Federal Radio Act of 1927,[2] later replaced by the Federal Communications Act of 1934.[3] Both statutes approached the problem of spectrum allocation in the same way by creating a regulatory agency to allocate portions of the spectrum to various uses and to issue a limited number of licenses for use of the spectrum thus allocated.[4] This federal regulation has been criticized for two reasons. First, the rights granted to licensees have been so inflexible that there are few incentives for them to make economically

1 *See, e.g.*, R. Coase, W. Meckling & J. Minasian, "Problems of Radio Frequency Allocation," May 1963 (unpublished manuscript at RAND Corporation, Santa Monica, Calif.); Coase, *The Federal Communications Commission,* 2 J. Law & Econ. 1 (1959); Gifford, *EMC Revisited—1966,* 8 IEEE Transactions on Electromagnetic Compatibility 123 (1966); Jones, *Use and Regulation of the Radio Spectrum: Report on a Conference,* 1968 Wash. U.L.Q. 71; Levin, *The Radio Spectrium Resource,* 11 J. Law & Econ. 433 (1968); Levin, *New Technology and the Old Regulation in Radio Spectrum Management,* 56 Am. Econ. Rev. Papers & Proceedings 339 (1966); Meckling, *Management of the Frequency Spectrum,* 1968 Wash. U.L.Q. 26.

2 Act of Feb. 23, 1927, ch. 169, 44 Stat. 1162.

3 Act of June 19, 1934, ch. 652, 48 Stat. 1064, 47 U.S.C. §§ 151 *et seq.* (1964). A history and analysis of this early regulatory period is given in Coase, *supra* note 1, at 1-7. The FCC's authority does not extend to federal government users, who are regulated by the President's Office of Telecommunications Management. 47 U.S.C. § 305 (1964).

4 A discussion of the FCC's present allocation procedures is found in Coase, *supra* note 1.

efficient use of communications resources—transmitters, receivers, and the radio spectrum itself.[5] Second, the regulatory procedures established over forty years ago have not been adapted to the dramatic changes in technology and demand that have occurred since the end of World War II.[6]

It has been suggested that the present FCC management system should be fundamentally altered by increasing the scope of private property rights in spectrum use, thus allowing the spectrum and related sources to be allocated more by market processes. Proponents of a market-oriented system have argued that private ownership of spectrum use, accompanied by few regulatory constraints, would permit spectrum owners to use the spectrum and communications equipment more efficiently—that is, in ways that more nearly satisfy consumers' preferences.[7] Others have argued that a market-oriented system would be unworkable due to technical characteristics peculiar to electromagnetic radiation—for example, the impossibility of confining radio transmissions to specific geographical spaces and to discrete bands of frequencies.[8] In short, it is feared that allocation by the market mechanism will bring back the confusion that prevailed before regulation in 1927.

It is our belief that the parties to this debate have never truly joined issue. The market mechanism *as a theory* cannot be offered as an alternative to FCC regulation; to make the debate useful it is first necessary to articulate a detailed system of property rights in spectrum usage.[9] Only after a property system has been defined

5 *See* R. Coase, W. Meckling & J. Minasian, *supra* note 1, at chs. II–IV; Jones, *supra* note 1, at 81–84, 97–107; Levin, *The Radio Spectrum Resource, supra* note 1, at 486–96; Meckling, *supra* note 1, at 26–27.

6 IEEE/EIA, Joint Technical Advisory Committee, Spectrum Engineering—The Key to Progress: A Report on Technical Policies and Procedures for Increased Radio Spectrum Utilization § III-C (1968); IEEE/EIA, Joint Technical Advisory Committee, Radio Spectrum Utilization—A Program for the Administration of the Radio Spectrum ch. 1 (1964); Telecommunication Science Panel of the Commerce Technical Advisory Board, U.S. Dep't of Commerce, Electromagnetic Spectrum Utilization—The Silent Crisis: A Report on Telecommunication Science and the Federal Government 1–18 (1966).

7 *See* sources cited in note 1 *supra*.

8 *See* D. MacQuivey, Comments on William Meckling's Paper on "Management of the Radio Spectrum," Sept. 11, 1967 (unpublished manuscript at Stanford Research Institute, Menlo Park, Calif.).

9 *See* Jones, *supra* note 1, at 85, 87, 114–15.

and tested can evaluation be made of the relative utility of the market system versus the regulated system now in effect.

The creation of a property system in the electromagnetic spectrum, upon which the operation of the market depends, necessarily involves three elements: (1) defining rights that are unambiguous and valuable but that are nevertheless compatible with the special physical properties of electromagnetic radiation; (2) devising a mechanism for legal enforcement of these rights; and (3) providing a means for initial distribution to the public of the spectrum-use rights now owned exclusively by the United States Government.

This essay will discuss the legal, economic, and engineering issues raised by these items and will propose a fully specified system of property rights for electromagnetic spectrum use. The proposed system should provide the basis for field experiments in certain portions of the spectrum, thus making possible an empirical test of the benefits and costs of alternative spectrum-management systems.

The essay is organized as follows: Chapter I presents a brief introduction to the physical and economic characteristics of electromagnetic radiation. The application of the general economic theory of property to the problem at hand is discussed in chapter II. Chapter III sets forth the proposed definition of spectrum-use rights, and chapter IV discusses the legal and transactional problems raised by the definition. Chapter V offers legal precedents and methods for disposal of government-owned resources, and outlines possible experiments for the transition from the present administered system to the market system.

I. Characteristics of Electromagnetic Radiation

A. Technical Dimensions of Radiation

Electromagnetic energy (radio and light waves) is transmitted in the form of oscillating electric and magnetic fields at the velocity of 300 million meters per second. Suitable control of the source generator for these fields makes possible the transmission of information without the use of wires or other physical connections. Communications by radio waves can be conveniently discussed in terms of three dimensions: the *time* during which transmissions occur, the geographical *area* over which the radio waves spread, and the *spectrum* or range of frequencies (also called the "band") used in the transmission.[10] These dimensions, taken together, will be hereafter referred to as "TAS" or the "TAS package."

Technically, the "spectrum" refers to the entire range of frequencies from zero cycles per second through billions of cycles per second. The propagation of radio waves is affected by the atmosphere, the ionosphere, the curved surface of the earth, and even mountains, buildings, fields, deserts, and oceans. Since these effects vary widely depending upon frequency, some ranges of frequencies may require different forms of property arrangements. This essay is concerned with a definition of property rights for only the portion of the frequency spectrum between 50 MHz and 1,000 MHz.[11] Aside from AM radio broadcasting and

10 In addition to frequency (measured in cycles per second, or "hertz"), other controllable parameters of electromagnetic waves are polarization, modulation, phase, and directional intensity.

11 MHz is the abbreviation for megahertz, or one million cycles per second. Among the uses currently allocated in this portion of the spectrum are:

54– 72 MHz	Television channels 2 to 4
76– 88	Television channels 5 and 6
88–108	FM radio broadcasting

microwave-relay transmission, the vast majority of electromagnetic communication presently takes places within this band of frequencies.

A radio communication system involves the transmission and reception of electromagnetic energy. By use of a transmitter and an antenna, energy is radiated into space. The energy is then intercepted by a receiving antenna, filtered to accept the wanted signals and reject the undesired ones, and processed to recover the information superimposed on the transmitted wave. The undesired radiation includes not only purposeful emissions from other sources but also casual or unintended radiation that is a part of our physical environment—caused by the sun, the galaxies, neon signs, automobile ignition systems, electrical machinery, and a myriad of other radiation sources.

To accomplish communication, the received field strength[12] of the desired signal in some combination of TAS must be greater, by a technically specified amount, than the combined strengths of all other signals in the same TAS dimensions. Two separate communications operators cannot use the same TAS package without interfering with each other's service. For example, two operators using the same combination of time and area must transmit on different frequencies to avoid significant radio interference; and if the time and frequency overlap, the two operators can avoid interfering with each other only by separating their transmitters and limiting the field strength of their signals.

A given level of communications service between two locations can be achieved with a variety of physical "input" combinations. A more powerful transmitter can be substituted for a higher antenna and still produce the same field strength at a given receiving location. Under certain conditions, greater power and a higher antenna can both be traded for a narrower band with the same

---

108–136	Aircraft navigation
174–216	Television channels 7 to 13
225–420	Primarily federal government uses
420–470	Amateur, citizens, domestic public, and industrial
470–890	UHF-television channels 14 to 83.

The complete allocation is in FCC Rules & Regulations § 2.106, 47 C.F.R. § 2.106 (1968).

12 Field strength (sometimes called "field intensity") at a given location is measured in volts per meter (v/m).

result. Determining the precise combination of productive inputs is an economic as well as an engineering problem, and the desirable combination will vary according to changes in technology, consumer demands, and relative prices for different technical components.

B. Signal Attenuation in Area and Frequency

Emitted radio signals cannot be confined or "fenced" in either area of frequency; signals travel indefinitely through free space, and radio transmissions on one frequency inevitably "spill over" to some extent into adjacent frequencies. For the purpose of defining property rights in spectrum use, however, neither problem is insurmountable.

Two factors influence the field strength of a radio signal in a particular geographic area. First, radio signals are generally obstructed or deflected by barriers such as natural topography or artificial structures. Second, signal field strength in free space diminishes with increasing distance between the transmitter and receiver; at sufficient distance from the transmitter, a signal's field strength becomes insignificant. However, the use of "free space" when calculating field strengths exaggerates the strength of radio signals traveling relatively near the earth's surface. The exaggeration occurs because the field strength of waves following the curvature of the earth attenuates much more rapidly than does the field strength of unobstructed waves.[13]

Analogous considerations apply in the frequency dimension. Due to practical limitations of electronic components, some energy is radiated outside the intended transmission band. Several precautions can be taken to reduce interference from such extraneous radiation. First, the rate of transmission and the method of modulation must be consistent with the time and bandwidth available within the TAS package.[14] Second, careful design of

13 *See* H. Reed & C. Russell, Ultra High Frequency Propagation 155 (1953); Norton, *The Calculation of Ground-Wave Field Intensity Over a Finitely Conducting Spherical Earth,* 29 Institute of Radio Engineers Proceedings 623 (1941).

14 A TAS operator can design his communication system in terms of several variables—duration of transmission, bandwidth, phase, and the ratio of the

transmitters can avoid the generation of spurious frequencies; proper filtering and shielding can reduce their transmission; careful design of receivers can greatly reduce sensitivity to such spurious signals. Third, adequate maintenance service for both transmitters and receivers can assure their proper functioning.

As noted earlier, successful communication does not require that the field strengths of unwanted radio signals—either from a different frequency or a different area—be zero at the receiver location, but only that they be a small fraction of the wanted signal. Technical methods are available for measuring, estimating, and controlling field strengths in both spatial and spectral dimensions. Such techniques are commonly used by the FCC to verify that its licensees conform to established standards for service and interference. In cases where a radio operator's field strength exceeds allowable limits, one or more of the technical inputs—such as antenna height or transmitter power—can be changed so as to modify the radiation pattern.

Not all variables affecting field strength are directly controllable, however. The rate of change of field strength with distance is a physical law that can be modified only by solid barriers such as mountains. Also, field strengths can be affected by the time of day, season of the year, and various ionospheric and meteorological conditions, some of which are random in nature. The impact of such effects on the problem of defining property rights in spectrum use will be discussed in chapter III, section B.

C. Economic Features of the Spectrum Resource

The phenomenon of interference makes spectrum use a scarce resource; is it not always available in sufficient quantities to allow

---

amplitude of the desired signal to the amplitude of noise plus interference (the "signal-to-noise ratio"). *Cf.* C. Shannon & W. Weaver, The Mathematical Theory of Communication 67 (1963).

Modulation is the process of varying one or more of these variables in accordance with the message to be sent, and thus provides "trade-offs" between time, bandwidth, and the signal-to-noise ration. For example, an ordinary television receiver produces a complete picture on the screen thirty times per second. In contrast, the Mariner-Mars satellite, using a much weaker transmitter and sending the signal a much greater distance, required five hundred minutes to produce a picture.

all users to have as much as they desire at zero price. In other words, any possible expansion in spectrum use would be valuable; individuals would be willing to sacrifice some amount of other goods or dollars to obtain additional TAS rights.

There are obvious manifestations of spectrum scarcity: In many areas of the United States, more people desire FCC licenses than can be satisfied given current equipment and interference standards; applicants for licenses may be assigned frequencies different from those requested; licensees are sometimes subjected to serious interference. In fact, the very existence of extensive FCC rules is an indication of scarcity; if there were no scarcity there would be no need for restrictions on who may use what frequencies where and no need for limitations on antenna location and height, transmitter power, and so on. Of course the degree of scarcity varies with different TAS combinations: VHF television is more valuable than UHF, and UHF is more valuable in New York City than in Butte, Montana. But all TAS combinations are scarce, and economic decisions must be made as long as the use of the spectrum by one person prevents or constrains its use by another or requires the other to incur a cost for additional or more elaborate equipment.

Thus scarcity imposes upon society the necessity of choice, a necessity that must be faced by any system of spectrum management, including a market system. The operation of a market in spectrum use depends, as we have noted, on the creation of a property system in the spectrum; that is, it depends upon the specification of a system of rights and duties of users sufficiently certain in meaning, sufficiently enforceable, and sufficiently valuable that people are willing to spend money to acquire the right. Several general criteria for defining property rights have been suggested in recent economic literature, and we turn to these general considerations in the next section.

II. Economic Theory Regarding Property Definitions, with Special Application to Spectrum Use

The economic theory of property rights is important to this study. Other studies have indicated that variations in the legal rules governing human action significantly influence economic behavior and the allocation of resources.[15] Different configurations of rights and duties will differently affect an individual's cost-reward calculus, providing him with varying incentives for different resource uses. A particular resource use can be more costly (or less rewarding) under one set of rights and duties than under another; and the more costly any given behavior is, the less of it we should expect to observe. For example, a person renting an automobile for one day obtains the right to drive the car, but not the right to sell it. The price of the right to drive depends on mileage, which is not a perfect proxy for wear and tear. Rough driving and neglect are less costly for the car renter than for the car owner, who must bear the cost of such actions in higher repair bills or lower resale prices. In most cases we would expect different behavior by the car renter and the car owner, stemming primarily from their different property rights.

In this section we discuss four general desiderata of property definitions found in the recent legal and economic literature. In each instance we note the consequences of different formulations of property rights and apply general theory to the problems involved in defining rights of electromagnetic spectrum use.

15 For discussions in the economic literature see J. Hirshleifer, J. De Haven & J. Milliman, Water Supply: Economics, Technology, and Policy 32–73, 222–54 (1960); A. Alchian, "Some Economics of Property," May 26, 1961 (unpublished manuscript, P-2316, at RAND Corporation, Santa Monica, Calif.); Alchian & Kessel, *Competition, Monopoly, and the Pursuit of Money,* in National Bureau of Economic Research, Aspects of Labor Economics 156 (1962); Coase, *The Problem of Social Cost,* 3 J. Law & Econ. 1 (1960); Demsetz, *Toward a Theory of*

A. Exclusivity

A resource is more likely to be allocated to its most valuable use when the right to its use is unambiguously exclusive—that is, when the right is assigned to one person (the owner) who controls all uses of the resource and who can exclude other persons from using it without his permission. The importance of exclusivity is best seen by analyzing the consequences of its absence.

Consider the case of a "common-pool" resource, such as fish in a lake or an oil and gas deposit, where exclusive rights to withdraw the commodity are not assigned and property ownership in the resource can be established only by capture. Since no fisherman can exclude others from fishing in the lake, none has an incentive to postpone a catch; the fish any fisherman fails to take now, in expectation of higher future value (larger fish), may well be caught by some other fisherman. Thus the total costs privately incurred by the fishermen do not equal the social costs of using the resource, since each fisherman ignores the future value of the fish. There is a social cost, in that the stock of fish is depleted at an uneconomically rapid rate.[16]

Property Rights, 57 AM. ECON. REV. PAPERS & PROCEEDINGS 347 (1967); Demsetz, *Some Aspects of Property Rights,* 9 J. LAW & ECON. 61 (1966); Demsetz, *The Exchange and Enforcement of Property Rights,* 7 J. LAW & ECON. 11 (1964); Gordon, *The Economic Theory of a Common-Property Resource: The Fishery,* 62 J. POL. ECON. 124 (1954).

Legal literature has not kept pace with the economic writings concerning the influence of legal rules on economic behavior. But a significant new literature is emerging. *See* Baxter, *The SST: From Watts to Harlem in Two Hours,* 21 STAN. L. REV. 1 (1968) (discussion of the inadequacy of present rules for allocating the social cost of sonic booms); Blum & Kalven, *Public Law Perspectives on a Private Law Problem—Auto Compensation Plans,* 31 U. CHI. L. REV. 641 (1964) (an investigation of the underlying rationale of tort liability and compensation schemes, with special reference to economic criteria for allocating accident costs); Calabresi, *The Decision for Accidents: An Approach to Nonfault Allocation of Costs,* 78 HARV. L. REV. 713 (1965) (formulation of a theoretical framework for nonfault liability in which risky activities reflect accident costs in market prices); Michelman, *Property, Utility, and Fairness: Comments on the Ethical Foundations of "Just Compensation" Law,* 80 HARV. L. REV. 1165 (1967) (an examination of the law of eminent domain in the light of economic and other criteria). *See also* Calabresi, *Some Thoughts on Risk Distribution and the Law of Torts,* 70 YALE L.J. 499 (1961).

16 *See* Gordon, *supra* note 15.

The same analysis can be applied with great force to oil and gas reservoirs, which prior to conservation legislation were subject to a legal "rule of capture."[17] While the right of occupancy of the land overlying the reservoir may have been exclusive, the right to remove hydrocarbons from the reservoir was nonexclusive. The more petroleum one man produced the more his neighbor had to produce to prevent the oil beneath his land from flowing to the first producer's land and out his well. Society incurred the cost not only of a resource too hastily depleted but also of resources put into unnecessary wells that sometimes lined surface boundaries like fenceposts.

There are two solutions to the nonexclusivity problem. First, excessive depletion can be prevented by establishing quota restrictions on the amount of the resource that each person may withdraw; in many instances this is done governmentally. A second solution is to establish exclusive ownership of the pool. The sole owner has a legally enforceable right to all the capitalized future values of using the resource because his ownership right allows him to withhold permission for other persons to use it unless he is compensated. With exclusive ownership, any uneconomical depletion of the resource becomes a relatively more costly form of behavior for the owner.

It was the nonexclusivity of spectrum-use rights during 1925–27 that initially produced government regulation by the Federal Radio Commission and later by the Federal Communications Commission. During this period an operator could transmit on whatever TAS dimensions he desired; this naturally led to severe interference problems in the preferred TAS combinations.[18] The FCC's solution was to grant exclusive rights to spectrum use in the form of licenses that permitted no flexibility in the uses to which rights were put and little flexibility in the mode of operation. The solution proposed in chapter III of this essay is to define exclusive property rights in spectrum use without adding economically inefficient restrictions.

17 *See* H. William & C. Meyers, Oil and Gas Law § 204.4 (1959).

18 This period, described as one of "chaos in broadcasting," Coase, *supra* note 1, at 5, was characterized by court decisions that prevented the secretary of commerce from limiting the number of licensees assigned to any combination of TAS. *See* Hoover v. Intercity Radio Co., 286 F. 1003 (D.C. Cir. 1923); United States v. Zenith Radio Corp., 12 F.2d 614 (N.D. Ill. 1926). As a result, more than 200 new

B. Costs of Exchange and Enforcement

Rights are more likely to be devoted to their most valuable use when they are freely exchangeable and divisible at low cost relative to their value. Legal restraints on the use and exchange of resources can prevent their owners from putting them to more productive uses. While flat prohibitions on alienation are rare in the law—in fact the Rule Against Restraints on Alienation and the Rule Against Perpetuities are designed to prevent clogs on alienability—there are instances of legal hindrances to transfer and subdivision. For example, it is unlawful to sublet rent-controlled apartments in New York City for a higher price. Hence the case of a law-abiding, elderly widow "trapped" in a twelve-room apartment. Similarly, the prohibition against the sale of reclamation project land to a farmer who already owned 160 acres of land in a reclamation project nearly stymied all sales of highly productive land in California's Central Valley.[19] The sale of water rights apart from the land has been prohibited in some Western states.[20] Analogous restrictions apply to FCC licenses for spectrum use. Certain "blocks" of frequencies are restricted to specified uses on a nationwide basis, allowing no exchange of licenses among different classes of use.[21] The spectrum property system proposed later in this essay avoids such restraints.

Exchanging rights is a costly process; it includes the costs for both buyers and sellers of searching out, negotiating, and enforcing mutually beneficial exchange opportunities. These costs can be affected by the way property rights are defined.[22] For example, vague and uncertain definitions of property rights will tend to re-

---

radio stations were established between July 1926 and February 1927, when the Federal Radio Commission was established. During this period "[t]hese stations used whatever power or wave length they wished, while many of the existing stations ceased to observe the conditions which the Secretary of Commerce had inserted in their licenses." Coase, *supra* at 5.

19 Meyers, Book Review, 77 Yale L.J. 1036, 1038–41 (1968).

20 *See* J. Hirshleifer, J. De Haven & J. Milliman, *supra* note 15, at 239–43.

21 *See* Federal Communications Act § 310(b), 47 U.S.C. § 310(b) (1964); FCC Rules & Regulations § 2.102, 47 C.F.R. § 2.102 (1968).

22 *See* Coase, *supra* note 15, at 15–19; Demsetz, *Toward a Theory of Property Rights, supra* note 15, at 354–57; Demsetz, *Some Aspects of Property Rights, supra* note 15, at 65–70.

strict exchanges by requiring greater negotiation and litigation costs. Also, exchange costs increase as more parties become involved in the transaction. The transfer of all or part of a resource right—whether between two parties in the same use, or from a given use to a more valuable use—will not occur unless the difference between the value in either instance exceeds the cost of making the exchange.

In general, the efficient property definition will be the one for which the costs of exchanging and enforcing property rights are smallest relative to the value of the rights. In the cast of property rights in electromagnetic spectrum use, the costs of negotiating and enforcing rights exchanges under a market system of spectrum management may be quite high relative to their value, due to difficulties in exchanging rights in the several TAS dimensions and in maintaining field strengths within area and frequency limitations. The exchange-cost implications of alternative spectrum-use property definitions are discussed in chapter III.

C. Externalities

Even though the right to a resource may be unambiguously and exclusively assigned and although it may be legally salable, its owner may use the resource in ways that have an adverse physical effect on the use of resources by others.[23] An example of a harmful external effect would be the famous case of an upstream factory polluting the water used by downstream farmers without first obtaining their consent. The case of a beekeeper's honey production being increased when a neighboring landowner plants an apple orchard provides an example of a beneficial external effect.[24] In each instance there is some social cost or benefit—often called an "externality"—which was not taken into account by the owner in deciding how to use his property rights; the total

23 The discussion given in this section is based primarily upon the work of Coase, *supra* note 15, at 2–19.

24 These two examples fall within the rubric of "physical" external effects, in which one owner's use of his rights alters the physical nature of other owners' property. Such effects are usually distinguished from so-called "valuation" external effects, where the action of one person affects the pecuniary value but not the physical character of another person's property. Consider the case of two com-

costs or benefits caused by his action do not equal the private cost he incurred or the benefit he obtained.

Externalities are produced either by the absence of legal rules to prevent their occurrence or by excessive enforcement or exchange costs where legal rules do exist. In the case of stream pollution, so long as the right to use the stream is clearly assigned to one of the alternative uses (for example, manufacturing), the other user (the farmer) will purchase it if he values the resource more highly in his use and if the costs of exchange and enforcement do not exceed the incremental value produced by the exchange. Even with zero exchange and enforcement costs, an exchange may not occur because the value of the use to which the right is first assigned (for example, manufacturing) is greater than the alternative use. The nature of the exchange will depend upon the initial assignment of the right; for example, if the factory has the right to pollute, the farmers might pay the factory to install purification equipment. Exchange and enforcement costs will increase, however, as the number of parties participating in the exchange increases.[25] The costs incurred in exchanging rights will

---

peting television broadcasters using the same time and area combination but different frequency bands. Frequency spillovers by broadcaster *A* into broadcaster *B*'s band would constitute a physical externality, while an improved programming schedule by *A* that lured viewers and advertisers away from *B* would constitute a valuation externality. This terminology blurs the issue somewhat since the ultimate effect in both cases is on the value of *B*'s right.

In discussing external effects this essay refers to the physical rather than the valuation variety. In the spectrum property system to be defined, we give existing TAS owners no protection against new competitors' superior products or lower prices. Following Demsetz's argument, *Some Aspects of Property Rights, supra* note 15, at 62–65, giving existing firms the right to retain their customers would require that new firms compensate them before going into business. This distribution of rights would discourage market entry by increasing exchange costs and uncertainty as a result of forcing newcomers to determine in advance of production which incumbent firms would be damaged by their competition and to what extent. Thus economic efficiency suggests that the right to entry be given to newcomers, leaving existing firms the options of lowering prices, increasing product quality, paying the new producers to cease production, or shifting their resources to their next best alternative use.

25 Several authors have suggested that this proposition is generally true for many resources, *see* sources cited in note 22 *supra,* although there are no specific empirical studies available to validate that point of view. The technology of transacting and enforcing spectrum use leads us to believe, along with R. Coase, W. Meckling & J. Minasian, *supra* note 1, at 103–06, that the proposition is true for this resource.

be less when there are one factory and one farmer than when there are one factory and one hundred farmers. In the case of a factory that pollutes the air over an entire city, the value of clean air to all residents will rarely exceed the costs of their getting together and purchasing either the factory or its right to emit smoke.[26]

To achieve an efficient allocation of resources, contracting between rights-holders with respect to externalities must be a practical possibility. The lower the costs of making and enforcing such a contract relative to the value of eliminating the externality, the greater the probability that resource rights will be put to their most valuable use and that the rights-holder will bear the total cost of using his rights (the ''internalization'' of external effects). As before, the rule for choosing among definitions of property rights is to reduce the costs of exchange and enforcement relative to the value of the rights. This rule has a corollary concerning the initial distribution of property rights: If it is known in advance that a given use (for example, factory waste disposal) is more valuable than another use (for example, farming) and if for reasons of policy or exchange costs an auction is considered impractical, then the rights should be assigned to the more valuable use. This assignment would eliminate the subsequent exchange costs that would be incurred if factory owners had to purchase rights initially assigned to farmers.[27]

Externalities in electromagnetic spectrum use can arise because of the difficulties of confining radio waves of a given field strength to a specified geographical area and to a designated range of frequencies. The property definition proposed in chapter III specificies the extent to which one spectrum user may physically affect others. Chapter IV specifies how spectrum owners under

26 In some cases regulation by the municipal police power might be the efficient solution. *See* Coase, *supra* note 15, at 17–18. Alternatively, it might be efficient for one individual to represent the group of people whose interests have been adversely affected—for example, in a class action. *See* Note, *The Cost-Internalization Case for Class Actions,* 21 STAN. L. REV. 383 (1969).

In urban areas a significant amount of electromagnetic interference is caused by the aggregate of individually small radiations due to automobile ignitions and neon signs. This is known as "spectrum pollution," and its most efficient solution might also involve use of the police power, perhaps through standards applied at the point of manufacture.

27 *See* Demsetz, *Some Aspects of Property Rights, supra* note 15, at 66.

the proposed property system can contract to change the initial field-strength limits.

D. Flexibility Versus Certainty

Defining a right essentially involves specifying what activities are permitted a rights-holder and what activities are prohibited. An activity can be placed in one of three definitional categories: (1) prohibited altogether; (2) permitted as part of the right; or (3) permitted if the consent of specified other parties is obtained. In the stream pollution example given earlier, a flat legal prohibition on the discharge of waste by the factory would put waste discharge in category 1; if waste discharge were put in category 2, the factory would be allowed to discharge into the stream; if the activity were put in category 3, each farmer would have the right to prevent discharge that affected his water rights.[28]

This classification scheme has two economic implications for defining property rights. First, activities should not be placed in category 1 unless the number of affected parties is so large that the total exchange costs effectively prevent rights exchanges. Second, if the transfer of rights involves identical exchange costs in categories 2 and 3, optimal resource allocation will result from either arrangement; if exchange costs differ, the activity in question should be placed in the category having lower costs. The primary problem in designing a market system is not to decide whether an activity should be placed in category 1 or category 2, but to define property rights so that it will be economically attractive to place an activity in either category 2 or category 3.

The specific problem at hand, however, is of broader scope than the stream pollution example. We must define spectrum-use rights for an entire system of rights-holders, all of whom will make essentially similar use of the resource and will simultaneously acquire similar rights and liabilities. This situation forces consideration of a larger set of rights possibilities than is involved in distributing a well-defined right to groups who will use it differ-

28 These categories of rights and actions, and an approach for choosing among them, are discussed in A. Alchian, *supra* note 15, at 18–20.

ently (for example, water use by factories versus use by farmers) or in defining rights for one individual given the extant rights of others. For example, placing an activity in category 2 provides each TAS owner with more flexibility in the use of his resource, for it removes restrictions on his rights with regard to that activity. Any single owner would benefit from such a classification, but if the comparable rights of *all* owners are simultaneously made more flexible, an individual owner may lose more from the exploitation of this flexibility by others than he gains from the exploitation of his own. Similarly, placing an activity in category 3 gives each TAS owner more certainty that his rights will not be physically affected by the activities of other owners,[29] but it also give each owner less flexibility to adopt new technology or to use the resource for different purposes.[30]

Our proposed definition of spectrum rights places an action in category 1 only when the number of parties whose consent should be required appears to significantly increase exchange costs relative to the value of exercising the right. This decision to limit the use of category 1 meets recent criticism of the inflexibility of FCC license privileges.[31] Moreover, so long as the parties who must consent to a transaction are appropriately specified, flat prohibitions on resource use (category 1) do not provide more protection from physically harmful activities than does a requirement that consent be obtained (category 3).

In some cases, our a priori information may be inadequate to support a clear choice between categories 2 and 3. Where this

29 Complete certainty in this regard is not necessary for the functioning of a market. Any landowner can cite many areas of uncertainty—for example, the introduction of commercial and industrial uses into a residential area may depress values—yet there is an active market in real estate.

30 Stated differently, each TAS owner will have less possibility of a decrease in the value of his rights due to the physical consequnces of the actions of others, but also less possibility of increasing his wealth through his own actions.

31 *See* note 5 *supra* and accompanying text.

The FCC's practice of prohibiting an action altogether (such as preventing spectrum rights from being used for purposes not expressly authorized by the license) reduces uncertainty for licensees, but also hinders the introduction of certain new spectrum technologies and effectively prevents the transfer of TAS privileges to new and more valuable uses. Prohibitions that may have been appropriate when first imposed tend to become increasingly inappropriate as time passes; even the creation of a mechanism for updating the prohibitions may suffer substantial lags.

choice does not appear to have a major effect on the economic efficiency of the system we have chosen category 2, preferring flexibility to certainty.

E. Criterion for Definition of Property Rights in the Spectrum

Neither the literature nor our analysis presents firm and unambiguous rules for formulating a property system that is economically efficient. Instead, as we have shown in this section, property-right theory demonstrates that definitions of property rights do affect the costs and gains of resource use. There are a variety of trade-offs to be considered; for example, less costly enforcement and greater certainty can be obtained at the expense of less flexibility and reduced transferability.

In formulating spectrum-use rights, our working rule is that the social value of production from the use of the electromagnetic spectrum should be maximized; we make choices that are expected to increase the value of the spectrum resource relative to exchange and enforcement costs. For example, while narrowly restricted rights are easier to enforce, they usually have a lower value because of the inflexibility inherent in that narrowness. Our application of property-right theory is necessarily tentative; it will not usually be possible to prove that our definition is superior to all alternatives. In certain cases, however, we can show that our proposed definition is superior to some alternatives in that it allows both greater flexibility and lower exchange and enforcement costs.

III. The Proposed Property System for Spectrum-Use Rights[32]

In this section we construct and analyze a "package" of rights for electromagnetic spectrum use between the frequencies of 50 and 1,000 MHz.[33] The rights are defined in section A in terms of the technical dimensions that give the resource economic value: the time of the transmission, the geographical area covered, and the portion of the spectrum over which radio waves are emitted. We discuss in section B some of the implications the proposed property system has for resource use. In section C we consider some alternative property definitions.

A. Dimensions of the Property Right

1. Time rights

The time during which a transmission occurs is the simplest dimension of electromagnetic radiation to specify, observe, and control. The time dimension of the spectrum-use right is defined as twenty-four hours per day in perpetuity. Subsequent arrangements for sharing the time dimension of a particular TAS package are left to private contracting. Time rights are transferable and divisable; the owner of a TAS combination has the legal right to transfer to any party or parties the right to use his property for all

32 The several property definitions considered in this section fall within a class of so-called "output" definitions first suggested in R. Coase, W. Meckling & J. Minasian, *supra* note 1, ch. 4. A draft of chapter III benefited from comments by Louis A. Rose, U.S. Department of Commerce.

33 The system of property rights designed for the frequencies between 50 and 1,000 MHz may not be technically suitable for frequencies below 50 MHz (such as those used for AM radio broadcasting) that are subject to ionospheric reflection and long-range propagation, and that are restricted by international agreements.

or any portion of the time period. Thus the time element might be divided into two parts, with *A* using the TAS combination from 12 midnight to 6 a.m. and *B* using it the remainder of the day. Or the entire package could be leased for eighteen months to *B*, with a reversion to *A* at the end of the period.

2. Area rights

For a specified time and band of frequencies the owner has two area rights under our proposal: (1) the exclusive right to originate radiation subject to the constraint that the field strength achieved by this radiation does not exceed a specified limit, expressed in volts per meter (*X* v/m), at any point outside his area, and (2) the right to be free, above the same field strength (*X* v/m), from radiation originating in any other area.

As defined by (1), a person having the right to originate radiation in a specified band of frequencies within area *A*, with a field strength not to exceed *X* v/m at or beyond the boundaries of his area, has at the same time a duty not to exceed *X* v/m at or beyond the boundaries of his area. Another TAS owner—*B*— has a right (a cause of action) against *A* only when *A*'s field strength exceeds *X* v/m in the specified band within *B*'s area. Similar rights and duties are defined for *A*'s area neighbor *B* in (2): *A* is protected against radiation of field strength greater than *X* v/m occuring in *A*'s area originated by *B* or any other spectrum owner. Thus, the rights and duties of *A* and *B* are symmetrical, and the initial field-strength limit of *X* v/m determines the level of interference between them.[34]

A hypothetical example of how areas of TAS packages might be defined in the Southern California region is shown in figure 1. Each TAS area has linear boundaries (the solid lines) on and beyond which the field strengths of the TAS owner must not exceed *X* v/m. However, the area boundaries for the Los Angeles owner,

34 The field-strength limits are stated as absolute levels, rather than as ratios of one signal to another. Ratios are commonly used in radio engineering because the communications achievable with specific equipment depend more on the ratio of the desired signal to the noise level than on the absolute level of the desired signal. However, rights defined in terms of a ratio would be circular and indeterminate, with each owner's rights depending on field strengths of other owners. In setting the initial field-strength limits, consideration should, of course, be given to the signal-to-interference and signal-to-noise ratios implied by various absolute levels.

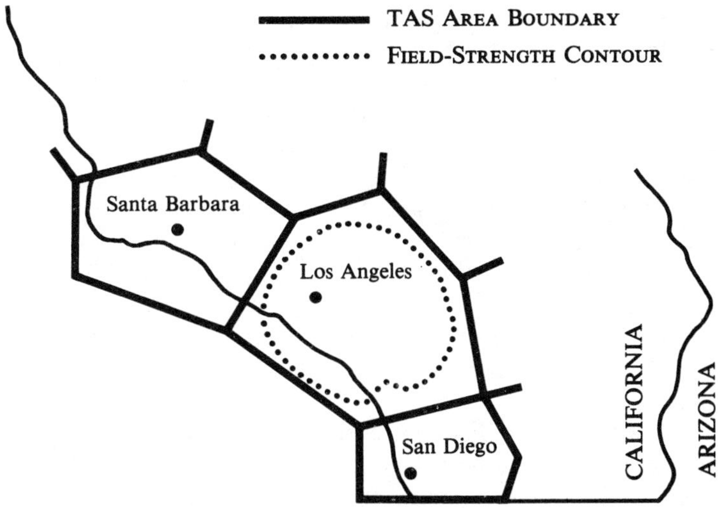

Figure 1. Hypothetical TAS areas in Southern California

for example, do not represent a "field-strength contour" along which field strengths would be equal to X v/m. This contour necessarily lies within the area boundary, as illustrated by the dotted line in figure 1, for the TAS owner would not be expected to achieve the maximum allowable field strength at all points (or even at any one point) on the linear boundary.

Defining TAS area boundaries in terms of straight lines rather than circular contours will reduce exchange and enforcement costs for two reasons. First, while circular areas would correspond to the spherical propagation of electromagnetic radiation, they would leave a no man's land where no owner would have exclusive legal rights to radiate; thus there would be no one from whom the right to service such areas could be purchased, should it become desirable. The uncertainties of squatters' rights in such unowned territories would increase exchange and enforcement costs, if indeed exchanges could be made at all. Second, straight-line boundaries simplify the combination and subdivision of TAS areas. Very substantial start-up costs are saved by using existing, well-understood conveyancing procedures instead of creating new procedures based on circles and usable only for a limited purpose. Rectangular or polygonal TAS areas may result in certain area "corners" receiving a signal below the authorized field strength, because the field strength achieved must not exceed X v/m at the boundary line closest to the transmitter. However, the proposed system of rights permits the TAS owner to service his corners by using directional antennas, by contracting with his neighbors to increase field strengths along existing boundaries, or by consensual redrawing of boundaries. Alternatively, a neighboring TAS owner could contract with the first owner to service the corner.[35]

The above specification of area rights will result in lower exchange and enforcement costs than would other possible rights definitions[36] because the proposed property system makes each TAS owner individually responsible for the excess field strength that he originates. Concomitantly, enforcement and exchange are facilitated because a TAS owner usually need deal with only one other such owner at a time. The total strength of unwanted signals permitted by these separate limits can be estimated from a knowledge of the number and location of adjacent TAS areas. In most cases the effects of one or two signals will dominate the calculation.

3. Spectrum rights

We noted in chapter I that radio signals spill over in the area dimension since they travel indefinitely through free space, and that they also spill over into other frequencies of the spectrum (so-called "out-of-band" emissions). Just as it was necessary to limit field strengths in the area dimension, it is necessary to limit them in the spectrum dimension.

The portion of the spectrum to which the owner's time and area rights apply is specified as a band of frequencies bounded by

35 Even though radiation is spherical, we place no restriction on field strengths in the height dimension. Although such a limit could be established, for example, at a height of 20 miles above some point within the TAS area in question, in practice there is no reason to do so, since at present there is no vertical separation of different communications services. However, when a ground-based receiver is near the transmitter's line-of-sight horizon, measured field strength increases sharply with the height of the receiver above the ground. Thus, to achieve greater specificity the definition of the right to radiate with a field strength of X v/m and of the right to be free of radiation from others in excess of X v/m could be further refined to provide that measurement of field strengths must be made at or below a given height above sea level.

36 *See* section C *infra* for alternative property definitions.

upper and lower limits. Within this frequency band the owner has two additional rights: (3) the exclusive right to originate radiation, subject to the constraint that the field strength achieved in any other unit band of frequencies not exceed a certain limit (*Y* v/m), and (4) the right to be free, above the same limit (*Y* v/m), from radiation originated by other sources assigned to a different band of frequencies.

Rights and duties for the spectrum dimension of the property right are thus symmetrical and are comparable to those for the area dimension. For example, within a given area the right specified in (3) gives *A* the right to originate radiation in area *A* over the band of, say 90.0 to 90.5 MHz, with a field strength not greater than *Y* v/m in another band, such as 90.5 to 91.0 MHz, owned by *C*. *C* has a cause of action against *A* only when *A* violates *C*'s property rights; that is, when *A*'s field strength in *C*'s band exceeds *Y* v/m in *C*'s area. Similar duties are defined in (4) for *A*'s frequency neighbors such as *C*: *A* is protected against any radiation having a field strength greater than *Y* v/m within *A*'s band and area originated by any other spectrum owner (such as *C*) using a different band. As in the case of area rights, the initial field-strength limit of *Y* v/m determines the permissible level of interference between TAS owners in the spectrum dimension.

Figure 2 illustrates the spectral density of two hypothetical signals and shows the application of the "out-of-band" radiation limit. *C* has a cause of action against *A* whenever the area under that portion of *A*'s field-strength spectral density function within *C*'s band (the shaded area)[37] exceeds *Y* v/m (the crosshatched area). *C*'s spillover into *A*'s band is subject to a similar limitation.

The *Y* v/m field-strength limit on out-of-band radiation is stated as an absolute limit rather than as a percentage of all radiation in the band.[38] In figure 2, *C*'s and *A*'s signals are both within *Y* v/m out-of-band constraint. Because *A*'s signal has much higher field strength than *C*'s signal, and because the out-of-band

37 This area is given by the integral with respect to frequency of *A*'s field-strength spectral density function, from 90.5 to 91.0 MHz. This integral (or area) corresponds to the total interference that *C* experiences due to *A*, since, in effect, a receiving antenna sums radiation over a limited bandwidth.

38 More certainty but less flexibility would be provided by specifying additional constraints on out-of-band radiation. For example, different limits could be set for subportions of the bandwidth in such a way that most out-of-band radiation would be confined to the edges of adjacent bandwidths.

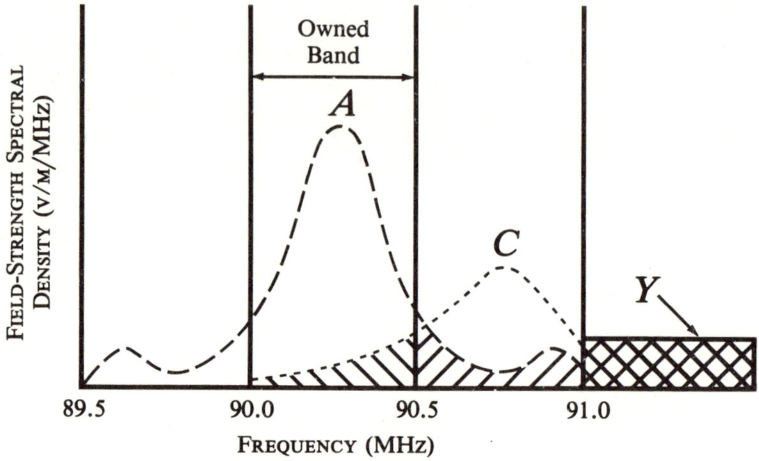

FIGURE 2. Hypothetical spectral density functions for the 90.0 to 90.5 and the 90.5 to 91.0 MHz bands

spillovers are approximately equal, *A*'s percentage spillover is much smaller than *C*'s. The absolute limit is advantageous to a TAS owner, such as *C*, who may wish to radiate a signal having lower field strength; if a percentage limit were used *C* might have to incur extra expense to obtain a transmitter with sufficiently low percentage spillover, thus eliminating the savings he obtains from purchasing a less powerful transmitter. Conversely, a percentage limit would allow *A* to create more spillover simply by using a more powerful transmitter; this would impose costs on *C*, who could either tolerate more interference or buy a more powerful transmitter himself.

While the spillover field strength is subject to an absolute limit, other spectral parameters remain unspecified. Thus, the choice of modulation method and polarization are left to the TAS owner. As long as the field-strength limits are not exceeded, the owner's selected modulation technique will not affect the amount of interference caused in neighboring areas or bands. Moreover, the owner may be able to widen the use of his TAS package by carefully selecting modulation and polarization techniques.[39] This

39 Allowing TAS owners to select modulation and polarization techniques, rather than specifying them in the TAS ownership certificate, is consistent with the "more flexibility, less certainty" option. Use of proper modulation and demodulation techniques allows more effective discrimination against interfering signals. Two

allows the proposed property system to achieve considerable flexibility in operation.

4. Usable TAS packages

Ideally, the boundaries and field-strength limits of the initial TAS packages would be economically optimal—that is, they would be perfectly suited (in each dimension) to the use that would result from existing marked and technological conditions. Initial optimality would save subsequent exchange costs that otherwise would be incurred in restructuring rights into optimal configurations. However, the information required for achieving initial optimality is not now available; obtaining the information could cost more than the resulting increment in value, and in any case this cost is likely to exceed the cost of achieving optimality through an initial round of exchanges. Some skillful guesses on original packaging should therefore be made, leaving the correction of mistakes to subsequent rights exchanges in the marketplace.[40] In addition, even if initially optimal TAS packages could be defined, they would not remain optimal over time. Changes in population concentrations, demands, costs, and technologies will necessarily require the restructuring of spectrum rights through exchanges. Thus, a viable property system for spectrum use demands only that the initial TAS packages be usable—that is, that the value of the rights exceeds the costs of enforcing them. Subse-

communications systems could operate without mutual interference within the same TAS package, provided that their modulation methods differed appropriately.

Similarly, under proper circumstances, two radio services, one using horizontal and the other vertical polarizations, could operate within the same TAS package. An electromagnetic wave consists of orthogonal electric and magnetic fields. A vertical linear antenna produces a vertical electric field, referred to as a vertically polarized wave. A horizontal linear antenna produces a horizontally polarized wave. In theory, a vertically polarized wave will induce no current in a horizontal receiving antenna, and vice versa. In the usual case, however, waves will be reflected, and reflection of a vertically (or horizontally) polarized wave can produce a horizontally (or vertically) polarized component. The result will be the production of a more general "elliptically polarized" wave, causing varying amounts of interference. If experience in a particular case shows that the level of interference is acceptable, transactions can expand the use of the TAS package without violating any other owner's rights. A consequence of this choice is that the field strength at the boundary will be measured with any or all polarizations.

40 This procedure has been followed in the disposal of other government-owned resources, such as land and water. *See* text accompanying notes 74–86 *infra.*

quent optimization through recombination of TAS packages may then be left to market processes.

An engineering study could establish preliminary values for the TAS dimensions. The time dimension has already been specified as twenty-four hours per day. The initial TAS area could be 70 to 75 miles across, centered on population concentrations. The unit band could be 25 thousand cycles per second (25 kHz) wide, a bandwidth suitable for voice transmission; hence the initial bandwidths of TAS packages would be multiples of 25 kHz. The field-strength limit for the area dimension (*X* v/m) should be large enough to overcome the ambient noise level caused by automobile ignition systems, electric motors, high voltage transmission lines and many other unintentional sources.[41] The limit on the strength of adjacent-band interference (*Y* v/m) should be on the order of one tenth of *X* v/m to result in a sufficiently low level of interference.

5. Summary

At this point it is possible to observe how the desiderata of property definitions developed in chapter II have been applied. First, the problems of exclusivity and externality are met directly by the definition of spectrum property rights. The exclusive assignment of rights for all three TAS dimensions to specific individuals or firms eliminates potential common-resource problems. The definition takes external effects into account by clearly limiting the extent to which any spectrum owner can spill radiation into a neighbor's TAS package. Second, no legal prohibition is placed on the transferability of spectrum-use rights or on the uses to which they are put. Third, we have avoided the complete prohibition of any activities by spectrum owners; instead, a symmetrical system of rights and duties is defined for each owner.

Finally, the proposed property system reflects certain decisions that should reduce subsequent exchange and enforcement costs. In most cases the reasoning behind each choice will not be

41 Sample median field strengths of ambient noise measured in a 25-kHz bandwidth (in microvolts per meter) for two frequencies are as follows:

TABLE 1

	50 MHz	1,000 MHz
Urban area	63	25
Suburban area	25	23

apparent until we discuss enforcement procedures and rules for exchange in chapter IV. However, one example that can be noted now is that our system of spectrum-use rights has been designed primarily for use by transmitter owners rather than for receiver owners. Not only would the former group—and especially commercial broadcast services—be more likely to acquire TAS rights, but the costs of exchanging rights would be less for this relatively small group than for the larger group of receiver owners.[42] Thus on both grounds,[43] TAS rights should be defined for originators of radiation.

B. Implications of the Definition

We seek now to demonstrate that the proposed property system in the spectrum deals effectively with the several problems of electromagnetic spectrum use caused by special technical characteristics of the resource.

1. Variation in propagation

We noted in chapter I that radio wave propagation is not entirely predictable. Achieved field strengths at any TAS location may vary with the time of day, the season of the year, and the weather. However, the property definition places absolute, nonstochastic limits upon the field strengths that a spectrum owner may achieve in other areas and bands. Any instance of excessive field strength amounts to a violation of duty owed to some other spectrum owner.

This imposition of absolute rights and duties is neither unrealistic nor inconsistent with the variability of wave propagation. First, the field strengths specified are maximum limits rather than requirements or predictions, and communications services can be designed with very high probability of operating within the limits. Second, some variations in propagation which are predictable by

42 There are hundreds of thousands more owners of radio and television receivers than owners of transmitters. Moreover, in cases such as the communications service operated by a taxi company, rights are defined for the entire communications service (including the dispatching transmitter) rather than for each individual cab (which contains both a transmitter and a receiver).

43 *See* sources cited in notes 27–28 *supra* and accompanying text.

time of day or season can be compensated for by adjusting transmitter power.[44] The nonpredictable component of the variation can be allowed for by operating below the field-strength limit. Third, the costs of going to court, the evidentiary requirements in court,[45] and the need for neighbors to get along with each other should ensure that small or infrequent violations will pass without protest.

One possible way to take explicit account of the unpredictable variations in field strength is to devise a stochastic definition for the spectrum-use property rights. For example, an operator could be permitted to exceed the field-strength limit (*X* v/m) in adjacent areas by up to 10 percent during 15 percent of the time he transmits. Such a stochastic definition of property rights would, however, significantly increase enforcement costs, since much greater monitoring activity would be needed to prove a violation. However, under the nonstochastic definition proposed in this essay, TAS owners could contractually provide for a stochastic definition to govern the question of interference among themselves.

2. Multipath propagation

A signal radiated from a point in one area may reach a point in another area both by line-of-sight propagation and by reflection from uneven terrain or buildings. This situation, known as multipath propagation, is depicted in figure 3. The signal radiated from point *T* in area *A* arrives at point *R* in area *B* by two separate paths: the line-of-sight path *TR* and the reflected path *TMR*, where point *M* represents the location of a mountain. As a result of multipath propagation, the combined field strength of *A*'s signals at point *R* in *B*'s area could exceed the *X* v/m limit, even though the field strength caused by each signal path alone does not. In such a situation *A* has violated his duty to *B* not to allow his field strength to exceed *X* v/m anywhere within *B*'s area. The property definition makes no exception for mode of propagation; if the total field strength at any point in *B*'s area exceeds the legally allowable *X* v/m, then *A* is liable. The fact that multipath

44 Such adjustments, while costly, are common for AM standard-broadcast licensees.

45 *See* chapter IV, section B, subsection 5 *infra*.

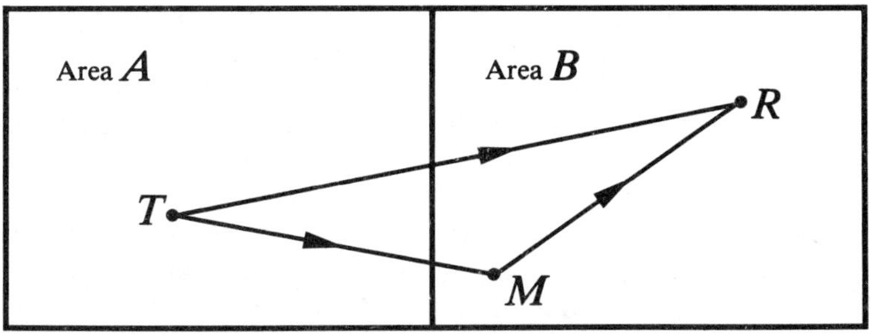

FIGURE 3. Multipath propagation of a radio wave from point *T* to point *R*

propagation is involved is irrelevant; spectrum owners are responsible for their own multipath propagation and are protected against that of others. When multipath propagation occurs or is thought to be likely, the owner can prevent it by using directional antennas.[46]

3. Intermodulation interference

The phenomenon of intermodulation has no close parallel in other resources. It occurs when radio signals transmitted on two different frequencies cause interference to an operator using the same time and area combination but a third, distinct frequency.

For example, suppose radio operators *D*, *E*, and *F* locate their transmitters on the same mountain and serve roughly the same area.[47] Assume that operator *D* transmits at a frequency of 100 MHz, operator *E* at 150 MHz, and operator *F* at 250 MHz. It is possible that *D*'s and *E*'s signals will combine via intermodulation to interfere with *F*'s signal even though *F*'s equipment is tuned to transmit and receive signals at 250 MHz only. In general, intermodulation interference occurs in a third frequency only when that frequency equals either the sum of the other two frequencies or their difference.

46 To reduce enforcement costs, this rule could be modified for multipath effects caused by transient reflectors such as aircraft, making TAS owners neither protected from nor responsible for this special case.

47 Obvious illustrations of such a situation are the television antennas on Mount Wilson, near Pasadena, Calif., or those on the Empire State Building, in New York City.

There are two basic physical processes through which intermodulation may occur, both of which are dealt with by the spectrum-use rights definition. First, intermodulation interference can arise from the proximity of *D*'s and *E*'s transmitters: *D*'s signals can affect *E*'s transmitter in such a way that *E* radiates energy not only at a frequency of 150 MHz but also at 250 MHz, thus interfering with *F*'s communications service. The proposed property definition assigns liability for this situation to *E*; so long as *E* has the right to transmit only at 150 MHz, his emissions at the frequency 250 MHz arising from intermodulation with *D*'s emissions are a breach of duty per se. It is left to *E* to correct this situation by choosing the most economical of several possible responses: (1) *E* could pay *F* to tolerate the intermodulation interference (which payment would not exceed the cost of any of the following options); (2) *E* could shift his operations to a fourth frequency that would not harm *F*'s or someone else's rights; (3) *E* could pay *F* to shift his operations to a different frequency; or (4) *E* could reduce *D*'s signal level inside his (*E*'s) transmitter, either by electrical isolation or by increased physical separation, in order to eliminate intermodulation products due to interaction with *D*'s signals. Since no one of these responses is likely to be best—in terms of cost versus value of reduced interference—in all cases of intermodulation, no particular response should be written into law. Rather, *E* should be allowed to find the most economical solution, subject of course to liability to *F* if no satisfactory solution is implemented.

The second process by which intermodulation can occur is where *D*'s and *E*'s signals combine to cause interference in *F*'s receivers but where, unlike the first case, such interference cannot be causally linked to either *D*'s or *E*'s transmitting equipment. Nevertheless, legal responsibility must be clearly assigned to avoid uncertainty of property rights that would increase litigation costs.

There are three rules by which liability for such interference could be assigned. First, the party who last began transmitting could be made liable for any intermodulation interference his radiation causes other operators. While this rule would increase certainty for preexisting operators, it might also lead to premature establishment of operations to avoid liability.

Second, *F* could be granted absolute protection against intermodulation interference, thus imposing on *D* and/or *E* the cost of eliminating the interference in the most economical manner. This might be done by (a) paying *F* to attach filters to his receivers to eliminate one of the two frequencies contributing to the interference, (b) paying *F* to shift his operation to a fourth frequency not subject to intermodulation interference, or (c) shifting either of their own operations to a frequency that would not harm *F*.

Third, *F* could be denied protection against this form of interference, thus forcing him to bear the cost of achieving an economic solution. *F* would make this choice by comparing the gains and costs of modifying his operation by filtering or changing frequency with those of paying *D* and/or *E* to change frequency. If the cost to *F* of attaching filters or of paying *D* or *E* to change his operation exceeded the value to *F* of using that particular frequency, the frequency would go unused.

For two reasons we choose not to protect TAS owners from the second form of intermodulation interference. First, the interference is generated in *F*'s receiving equipment and is easily controllable there by *F*. Second, to protect *F* against this intermodulation process would involve difficulties in apportioning liability between *D* and *E*.[48] This could lead to increased litigation costs. Thus, because of technology and costs of enforcement, we do not give spectrum owners protection against intermodulation interference generated in their receivers. Since the rights definition grants no protection, it implicitly denies it, for the definition is intended to be complete.

4. *Factors affecting the size of the service area*

We have already noted that the service area within which a TAS owner can achieve communication of acceptable quality will rarely extend to the boundary of his defined TAS area. The exact size of his service area is a function not only of his TAS rights and his communications operations, but also of the rights and operations of neighboring owners. This subsection explores some of the technological constraints and economic incentives that affect service area size.

48 Similar liability-allocation problems arise when the sonic booms of two supersonic jet transports combine to produce an intensified shock wave. Such waves can cause substantial ground damage, even though the sonic boom of each aircraft taken alone would not have been strong enough to produce significant harm. For a discussion of this problem, with reference to applicable doctrines of tort law, see Baxter, *supra* note 15, at 44–46 & n. 60.

Boundary effects and service area. The usefulness of a radio signal is not dependent upon the magnitude of the signal alone. Of more practical importance is the ratio of the magnitude of the desired signal to the magnitude of interfering signals. Communications will be poorest where the signal-to-interference ratio is smallest, and this will occur near the boundaries between neighboring TAS areas.

Suppose that TAS owners *A* and *B*, operating at the same time and in the same band of frequencies, share a common boundary and that each achieves a field strength of *X* v/m—the legally allowed limit—at that boundary. A receiver at the boundary would receive both signals equally well; the signal-to-interference ratio would be one, and the two signals would thus strongly interfere with each other. If the receiver is moved away from the boundary and toward the center of area *A*, signal *A* will grow stronger and signal *B* will grow weaker; that is, the signal-to-interference ratio will grow larger as the receiver approaches transmitter *A*, and the quality of the reception will improve correspondingly. The receiver may have to be moved a substantial distance back from the boundary, however, before the signal-to-interference ratio increases sufficiently for the reception to be of acceptable quality. Thus, there is a strip of territory, enclosing the *AB* boundary on both sides, within which mutual interference between *A*'s and *B*'s radiation can preclude satisfactory reception of either.

The effects of this strip of low- or uncertain-quality service area can be ameliorated in two ways. First, interference between two stations on the same frequency will be objectionable only when both stations are transmitting simultaneously. While some radio services, such as commercial broadcast stations, require continuous operation, the majority of radio services transmit only intermittently; in these cases, where only one operator is transmitting at a time, the area of acceptable-quality service will extend to the TAS boundary. Second, the width of the low-quality service strip is greater for transmissions at lower frequencies than at higher ones.[49] Thus, to keep the low-quality service strip a small fraction of the total area, the lower frequencies could be used for

[49] Table 2, below, demonstrates that the low-quality service strip will be wider for signals of lower frequencies than for signals of higher frequencies. Table 2 shows the distances from the boundary—the "setback distances"—at which signal-to-interference ratios of 100:1 and 1,000:1 can be achieved at the location of the receiver using modern frequency-modulation equipment. A signal-to-interference

transmitters serving large areas, while the higher frequencies could be used for transmitters serving small areas.

In general, however, a receiver must be set back some fixed distance from the *AB* boundary if a given level of immunity from interference is required for acceptable reception. The size of the setback distance—and thus the width of the low-quality service strip—will depend upon the level of immunity required and upon the frequency band of the transmissions. Since the setback distance is fixed, the percentage of the TAS area comprising acceptable-quality service area depends upon the size of the TAS area;[50] for a given TAS area, the percentage that is serviceable decreases as the setback distance increases. Table 3 shows the ratio of acceptable-quality service area to a calculated, circular area for two signal-to-interference ratios, three frequencies, and four circular areas.[51] For example, consider a circular area with a 25-mile radius: for operation at a frequency of 50 MHz with a minimum signal-to-interference ratio of 100:1, the service area will cover 62 percent of the circular area. At 1,000 MHz and a minimum signal-to-interference ratio of 100:1, service is provided to 92 percent of a circular area with a 50-mile radius. Thus table 3 demonstrates that interference-free service need not be purchased at the expense

ratio of 100:1 provides excellent voice communication, and a ratio of 1,000:1 provides high-quality service for nearly any type of communication. In the calculations it has been assumed that the radio horizon is within the boundary of the TAS area. *See* text accompanying notes 53–55 *infra*.

TABLE 2. DISTANCE FROM AREA BOUNDARIES AT WHICH TWO DIFFERENT SIGNAL-TO-INTERFERENCE RATIOS ARE ACHIEVED

Frequency	50 MHz	400 MHz	1,000 MHz
Distance for 100:1 signal-to-interference ratio (in miles)	5.3	2.7	2.0
Distance for 1,000:1 signal-to-interference ratio (in miles)	14.0	7.0	5.2

50 Even with zero setback distance, the serviceable area of a polygonal TAS area will ordinarily be less than the TAS area. This is true because the limit of X v/m field strength at or beyond any boundary of a TAS area limits the radius of the field-strength contour to the distance from the transmitter to the nearest boundary. Thus, if the TAS area is a square and if the field-strength contour of X v/m is a circle tangent at four points to the square (which implies the use of a nondirectional antenna), the maximum service area, with zero setback distances, will be 78.5 percent of the TAS area.

51 The circular areas used in table 3 represent the areas within a field-strength contour of X v/m. Thus, in table 3, the actual TAS area might be a square measuring 25 miles on each side; the radius of the circular area, which could be inscribed in this TAS square, would be 12.5 miles (representing the field-strength contour within the TAS area); and the percentage of acceptable-quality service area, for a

of large, low-quality service strips between adjacent service areas. Under a market system of spectrum use, moreover, those strips that do remain need not stay permanently unserviced. If the strips are regions of economic value—for example, if they contain substantial population concentrations—they will probably be transferred through market transactions to an operator who can and will service them; if the strips have no significant economic value, they will be of little interest in any case.

TABLE 3. PERCENTAGE OF CIRCULAR AREA THAT IS ACCEPTABLE-QUALITY SERVICE AREA[52]

Frequency	50 MHz		400 MHz		1,000 MHz	
Signal-to-interference ratio . .	100:1	1,000:1	100:1	1,000:1	100:1	1,000:1
Radius of circular area:						
12.5 miles	33	1	61	19	70	34
25.0 miles	62	19	79	52	84	63
37.5 miles	74	39	86	66	90	74
50.0 miles	80	52	89	74	92	80

In accomodating a particular radio service, there are choices open to a TAS owner other than those shown in table 3. For example, there are economies in using low frequencies for longer distance communications: lower antenna height and lower transmitter power are thereby permitted. The design of a particular communications system will take into account these additional factors.

Radio-horizon effects and antenna height. A strip of low-quality service area along both sides of the *AB* boundary arises when the signals from *A* and *B* are of comparable strength at the boundary. The width of the strip will depend upon the rate at

---

1,000-MHz band at a signal-to-interference ratio of 100:1, would be 70 percent of the area enclosed by a field-strength contour of *X* v/m whose radius is 12.5 miles.

The use of circular areas for the purpose of table 3 is not inconsistent with our earlier proposal to make TAS areas polygonal with straight-line sides. The purpose of table 3 is to show what portion of the area within the legally permissible field-strength contour of *X* v/m will be acceptable-quality service area under various conditions.

52 The numbers obtained in table 3 for the fraction of the circular area that is usable service area were calculated on the assumption that transmitting antenna heights were such that their horizons fell inside their respective TAS areas. A discussion of the effects of the radio horizon and antenna heights is given in the following subsection.

which *A*'s signal attenuates as it travels into *B*'s area and vice versa. The more rapid the attenuation of *A*'s signal, the more rapidly the signal-to-interference ratio will increase as a receiver is moved toward the center of *B*'s area, and the more rapidly the reception of *B*'s signal will improve. The more rapidly the reception improves, the greater will be the fraction of *B*'s area in which service is of acceptable quality.

The attenuation rate of electromagnetic radiation originating at an earthbound antenna depends upon whether the radiation is traveling on a line-of-sight path from the antenna or in a "shadow zone" of the antenna.[53] Assume, for simplicity, that the earth is uniformly curved and smooth, without mountains or buildings. A transmitting antenna will have a horizon whose distance from the antenna depends upon the height of the antenna: the higher the antenna, the more distant its horizon.[54] (An exact analogy is found in the observer who climbs a tower to be able to see farther.) The geometry of propagation of earth-originating electromagnetic radiation is shown in figure 4. Radiation from the antenna will not abruptly terminate at the horizon; rather, there will be some that propagates into the shadow zone beyond the horizon. The important fact for this discussion is that radiation propagating in a shadow zone attenuates much more rapidly with distance than does radiation traveling on a line-of-sight path from the transmitting antenna. The presence of natural and man-made obstacles on the earth's surface does not affect this conclusion. It does greatly complicate calculations of actual radiation patterns

53 Radio waves originating from a source in free space travel in straight lines. The electromagnetic energy spreads uniformly in all directions, and the field strength is at all points inversely proportional to distance from the source. Free-space propagation can be used for calculations of radiation between artificial satellites and as an idealization suitable for simple calculations of radiation patterns on the earth. It is, however, seldom an appropriate model for calculations involving practical communications problems, because energy is not radiated uniformly in all directions from earthbound antennas. The radiation paths are bent by refraction in the atmosphere or are reflected from the ionosphere or from solid objects such as buildings, hills, mountains, and the earth's surface. The combination of these refractive and reflective phenomena gives rise to the characteristics of electromagnetic propagation in shadow zones.

54 If an antenna has a height of h feet, the approximate distance d (in miles) to its horizon is given by the formula $d = \sqrt{2H}$. Thus a 200-foot antenna has a 20-mile horizon, an 800-foot antenna has a 40-mile horizon, and a 5,000-foot antenna only a 100-mile horizon.

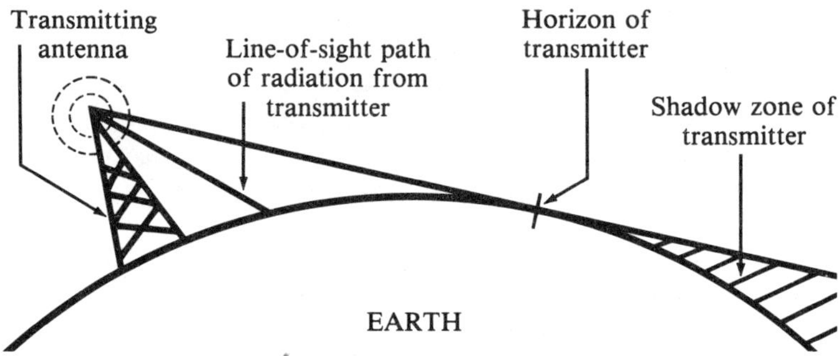

FIGURE 4. Geometry of radio-horizon effects

by obstructing line-of-sight paths and by creating, in effect, numerous small shadow zones. In most cases, however, only the earth's curvature can be relied upon to restrict signal propagation.

The importance of this radio-horizon effect for reducing the size of the low-quality service strips arises from the lack of any restriction on transmitter antenna height in the proposed property definition. *A* could permissibly construct an antenna sufficiently tall that there would be line-of-sight radiation paths from it into *B*'s area, and thus a wider low-quality service strip on *B*'s side of the *AB* boundary. In all cases, therefore, it is desirable for *A*'s antenna to be sufficiently short that *A*'s radio horizon falls within his own TAS area, since that will enlarge *B*'s service area.

The appendix at the end of this essay explains how TAS owners have independent incentives to keep their antennas low enough to meet the above condition. Generally, the *X* v/m limit at the boundary can be met with various combinations of antenna height and transmitter power. A low antenna that places the radio horizon within *A*'s area combined with a high transmitter power will result in higher field strengths (and hence larger or higher-quality service areas) within *A*'s TAS area than will a high antenna and low power combination that also meets the *X* v/m limit. On the other hand, a taller antenna will not improve the field strength within *A*'s area when the area already lies completely within the horizon, and will substantially increase the costs of *A*'s communications equipment.[55]

[55] These incentives would be muted if increased antenna height were nearly free, as it might be over a certain range due to the presence of nearby mountains or tall

The lack of any restrictions on antenna height in the proposed property definition does not, therefore, encourage the construction of undesirably tall antennas. On the contrary, physical and economic constraints provide implicit incentives for TAS owners to keep their antennas sufficiently short so that their horizons fall within their own TAS areas—that is, TAS owners have incentives to benefit their neighboring owners rather than to harm them. By so doing, they will maximize their own service areas without diminishing those of neighboring TAS owners.

Modification of the property right. We have suggested that TAS owners have economic and technical incentives to exercise their rights in ways that do not reduce the size of adjacent service areas, and that these incentives can be achieved within a property system that does not sacrifice flexibility for certainty. Similar results could be achieved with correspondingly less flexibility and more certainty: The property right could be modified to require that the radio horizon of each TAS owner lie within the boundaries of his TAS area.[56] By limiting the distance between the transmitter and the nearest area boundary, the rate at which *A*'s signal field strength attenuated beyond his area boundaries would be constrained. Consent of area-adjacent owner *B* (using the same time and spectrum combination) would be required before *A* could extend his radio horizon into *B*'s area.

C. Alternative Property Definitions

The property definition outlined above does not represent the only method of structuring a market system of spectrum management. The criteria of legal exclusivity and transferability discussed in chapter II can also be achieved by specifying field-strength

buildings. Note, however, that even the Empire State Building (1,472 feet high) has a horizon of about 54 miles, and that a 5,000-foot antenna is necessary for a 100-mile horizon.

56 The radio horizon distance d (in miles) could be defined as $d = \sqrt{2H_t} + \sqrt{2H_r} + \Delta$, where H_t is the height (in feet) of the transmitting antenna, H_r is the height (in feet) of the receiving antenna, and Δ is a function of frequency and ground conductivity (for example, seawater versus land). This equation could be further modified to permit, as a matter of right, directional antennas that radiate little in the direction of the near boundary, and to account for natural obstacles that limit the field strength at the boundaries.

limits in a slightly different manner. Although we believe that the definition given above is the preferable one, the three alternative definitions discussed below were at one time considered. In the following discussion we have labeled the definition given in section A as the positive-limit definition. Differences in exchange and enforcement costs among these definitions will be discussed in chapter IV, section B, subsection 1.

1. The zero-limit definition

This definition differs substantively from the positive-limit definition in only one respect; rather than setting a positive field-strength constraint of *X* v/m for radiation spillover into adjacent TAS areas, this definition sets a field-strength constraint of zero v/m outside the area. In practical terms, it would be impossible for any originator of radiation to achieve a field strength of zero v/m outside his TAS area, because radiation continues indefinitely through space at some positive field strength. Thus, while it may be convenient to speak of a "right" to achieve field strengths of zero v/m outside an area, such a statement is physically inaccurate as well as a contradiction of legal terms. A more accurate description of such a definition is a right to originate a signal of such strength that other owners are not caused undue harm. (A legal analogy would be the law of nuisance.) The result would be that courts would determine the limit of *X* v/m by ad hoc adjudication.[57]

2. The residue definition

In contrast with the positive-limit definition which made a TAS owner responsible for his emissions exceeding *X* v/m at or beyond the boundaries of his area, the residue definition makes the TAS owner responsible for all radiation above the limit *Z* v/m passing from his TAS area into immediately adjacent TAS areas, whether the radiation is originated by him or by some other TAS owner. For example, in figure 5, *B* has the legal responsibility for all

[57] The principle is the same for out-of-band emissions. The "right" to emit out-of-band emissions could be fixed at zero, with liability for breach of duty dependent on proof of harm by an objecting TAS owner in an affected band. The difficulties with such a definition of the property right in the spectrum dimension are exactly the same as those in the area dimension.

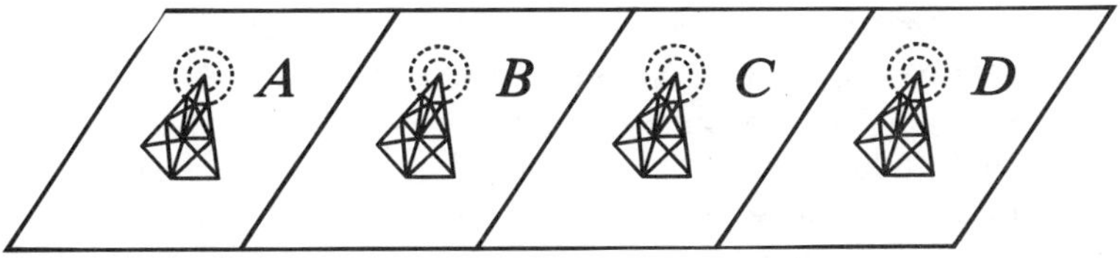

FIGURE 5. The areas of TAS owners *A, B, C,* and *D,* all of whom use the same time and spectrum combination

radiation reaching the *AB* boundary, even if some of that radiation was originated by *C* and *D.* Thus, if *Z* v/m were set as the maximum field strength of all radiation at the *AB* boundary and if measured field strength exceeded *Z* v/m, *B* would be required either to pay damages or to reduce his radiation to the point that total field strength at the boundary does not exceed *Z* v/m. *B*'s right to radiate in his TAS area thus depends on the amount of radiation reaching his area from other TAS areas, and the right is thus denominated a residual right. Under the residue definition, the same duty obtains for each TAS owner at each of his boundaries.

Clearly, *B*'s right to originate radiation is less certain under the residue definition than under the positive-limit definition to the extent that *B* is unable to predict radiation levels and attentuation rates of signals of other TAS owners.[58] It should also be noted that the field-strength constraint of *Z* v/m will necessarily exceed the field-strength constraint of *X* v/m for the positive-limit definition, because the residue field-strength limit must account for the total radiation leaving *B*'s area, and not just that originated by *B.*

3. The boundary–positive-limit definition

This definition is a modification of the right specified in the positive-limit definition. The TAS owner of area *A* (for a given

58 Uncertainty is not unique to this definition of the spectrum-use right. Under the positive-limit definition a TAS owner is protected against every other TAS owner individually, but he may not be able to predict what the total radiation from others will be. Since the performance of a communications system depends on the ratio of the desired signal to all interfering radiation, neither definition guarantees a given level of communications. Under the positive-limit definition the desired signal is guaranteed, but not the level of interference; under the residue definition the situation is reversed.

time and spectrum combination) would be granted (1) the exclusive right to originate radiation subject to the constraint that the field strength achieved by this radiation must not exceed *X* v/m at any point on his area boundaries, and (2) the right to be free from radiation exceeding *X* v/m at the boundary with each other TAS owner (such as *B* or *C*). In short, under this definition field strengths are measured only at the boundaries of each TAS area and not, as in the case of the positive-limit definition, at any point outside the TAS owner's area.[59]

[59] Analagous boundary rights could be defined for the spectrum dimension in terms of the field-strength density functions shown in figure 2. *A* could be required to meet a constraint on his density function where it crossed the lower boundary (90.0 MHz) and the upper boundary (90.5 MHz) of his bandwidth. However, we need not consider the spectrum dimension of this definition in detail because it suffers the same defects as the area dimension, as noted in chapter IV, section B, *infra.*

IV. Legal and Transactional Aspects of the Proposed System

Having set forth a proposed system of property in the spectrum, we turn now to the legal characteristics of that system. We consider in section A some general rules that should be incorporated into a statute if market processes are to allocate resources efficiently. Then in section B we deal with enforcement problems. These run the gamut from the effect of the legal definitions on enforcement and exchange costs to questions of who may enforce the right, who is liable, what tribunal shall have jurisdiction, and what remedies are appropriate.

A. General Statutory Rules

The statute should state that the general purpose of the property system and of the experiments outlined in chapter V is to create property interests in a designated portion of the spectrum in order to promote market allocation of this resource. Courts and and administrative agencies should be directed to interpret the statute in accordance with this objective. The statute should define the basic property interest in the spectrum and should delegate authority to a federal agency (presumably the FCC if its opposition to the system does not disqualify it from acting) to create TAS packages for sale to the public. The statute should declare generally that TAS packages are private property and freely transferable, and should specifically state that: (a) TAS owners may agree to changes in the boundaries of TAS areas; (b) TAS owners may transfer their rights to radiate on a time basis, by the hour, day, month, or year or by any other division of time; (c) TAS owners may agree to an increase or decrease of signal field-strength limits; (d) TAS owners may transfer all or any part of the band assigned to them; (e) TAS owners may transfer all or any portion of their geographical areas of operation; and (f) there shall be no restrictions, either in the TAS owner's certificate of ownership or by regulatory order, on transmitter power, or on antenna location and height. While the TAS owner is legally obligated not to radiate outside his geographical area or frequency band in excess of the field-strength levels specified in the certificate, the mix of inputs to achieve the permissible levels of field strength is a matter for decision by the TAS owner.[60]

The statute should require that all agreements by TAS owners relating to the use of the spectrum be recorded in a central registry. This requirement would cover changes in boundaries of TAS areas, sales or leases of TAS combinations or of any TAS component, and any other transaction affecting the use of a TAS combination that a potential purchaser would care to know about. A federal agency should be charged with the duty of maintaining these records for public access. Penalties should be assessed for noncompliance with the registration requirement, and unrecorded transfers should be voidable at the instance of subsequent purchasers.

The operation of a market in TAS rights will be greatly facilitated by a government-operated records office in which the status of all rights to radiate in the country may be readily ascertained by reference to indexes of bands and geographical areas. In short, what we propose is a kind of tract index for TAS combinations, like those maintained by land title companies. A closer analogy is the record-keping operation of state water engineers, who keep files in a central office on all water-use licenses issued by the state. The absence of easily located records greatly impaired water transfers and capital investment in water projects during the nineteenth century because investors did not know who owned the rights and what the rights consisted of. In order to plan a new water project, an investor had to make his own survey of uses on a stream and then ascertain, so far as possible, what each user's legal rights were in respect to quantity, time of use, and priority of right. Obviously the proposed property system should avoid this deterrent to the operation of the market.

60 If greater certainty and less flexibility appear desirable—as in the case of a limit on radio horizons (*see* appendix *infra*)—provision (f) of the statute will have to be modified accordingly.

The statute must stipulate the time duration of the rights created in the spectrum. We prefer creation of perpetual rights in the spectrum, corresponding to the fee simple absolute. The market tends to operate more efficiently when the time-tenure of the property interest is of long duration, since predictions about the usable life of specialized capital investments are thereby made less critical. We recognize, however, that there may be strong objections to this proposal, on at least two grounds: (1) The proposal is an experiment and if it fails the government ought to be able to recapture the resource without resort to eminent domain. We believe, however, that the advantage of free exchange outweighs the advantage of ease of recapture. (2) By selling the TAS combinations for a perpetual time period the government cannot benefit from the increase in value of the rights over time—that is, public wealth is transferred to private pockets.

There are several difficulties with this argument. First, it assumes the answer by claiming that the spectrum ought to be government property. While it is true that the government appropriated the resource in 1927, it did so not on the ground that the government was entitled to the wealth created by use of the resource, but rather on the ground that regulation was necessary for the resource to be useful at all. No doubt official opinion now regards the spectrum as government property, but the justification for this opinion is not self-evident. Second, the argument assumes that the price paid for the resource upon its initial distribution to the public will not reflect its discounted future value. This may or may not prove to be true, but we know of no persuasive evidence that the government has failed to receive the discounted future value of other resources, such as offshore oil exploration rights, that have also been sold by auction. Third, the fact that the government does not share in the unanticipated increase in the value of the resource has no bearing on economic efficiency. If the highest bidder obtains the resource at the initial auction, the fact that he earns an economic rent by a later sale to someone with a more profitable use of the resource means only that efficiency of resource use has been increased. "Rent" returns from an increase in value of the spectrum need not present income distribution questions any different from those presented by value increases of other resources. If the progressive income tax is the appropriate device for achieving redistribution of rent returns from other resources, it seems equally appropriate for the redistribution of rent returns from the spectrum.

However, since many people will object to government loss of unanticipated rents, an alternative system can be employed: The government could lease the TAS packages for a specified length of time, the duration of which would be sufficient for any potential investor to recover his investment during the term of the lease; where there is doubt about the length of the required period, it would be better to err on the long side. To further protect the TAS owner, the auction for re-lease of the TAS package at the end of the lease term should be set for a time somewhat earlier than the expiration of the term. For example, if *A* owns a TAS package with a term of fifteen years, the auction for re-lease of the package might be held in the tenth year, giving *A* (if he is not the successful bidder at that time) five years to phase out his operation.

This lease arrangement might also be accompanied by a system of rental payments instead of a lump sum payment due at the beginning of the term. The rental system may facilitate market entry, but it does complicate the collection process. Should a rental system be adopted, the statute or regulations thereunder should specify the remedies available to the government for default in rental payments by a TAS lessee. Such a provision might read as follows:

> Lessee shall pay as rent for the use of the TAS combination herein specified the sum of $________ (the amount of the bid) per year, payable semiannually in advance on January 15 and July 15 of each year. In the event of default on the payment of any rental installment when due, the United States of America may terminate this lease, and all the rights and privileges hereunder shall ipso facto cease; or the United States of America may, at its option, recover from the lessee all rent due and unpaid at the time of suit and may also recover all rent remaining to be paid for the balance of the term of the lease.

This provision gives the government the option of terminating the lease and re-leasing the TAS package or of standing on the contract and collecting all rent due under the lease; the second option is in effect an acceleration clause. In addition, it may be desirable to give the government the right to terminate the defaulting lessee's interest and to re-lease the TAS package, holding the defaulting lessee liable for the difference between the rental obtained under the new lease and the rental due under the old lease.

All of these remedies have traditionally been available to landlords for default on rental payments by a tenant of an estate for years in real property.

The principal problems that arise from defaults in rental payments are the liability of transferees for rental payments and the effect on their rights of the original lessee's default. Since the right in the spectrum can be divided in so many ways (time, area, frequency band, and field strength), apportionment of the rent among assignees is impractical. Thus, except where the full TAS combination is transferred, the original lessee should have the liability for the rent. Transferees of the original lessee will be subject to the forfeiture provisions of the lease and will have to protect themselves by contract, as for example, by requiring advance notice of the payment of rent by the lessee, with an opportunity to pay it themselves if the lessee defaults and with indemnification rights against the lessee. The legal situation described here is like that between a lessor of land for a term of years and a subtenant: The original lessee is liable for the rent, and the subtenant has no duty to pay rent to the lessor.[61] The subtenant is, however, liable to eviction either by statute or by standard lease clause, if the original lessee defaults in paying rent, since the termination of the lessee's estate terminates the subtenant's estate. The protection of the subtenant against such termination is left to private ordering.

While some may think that landlord and tenant law that originated in the sixteenth century is hardly appropriate for determining rights to spectrum use in the twentieth century, *some* method of enforcing the government's rent claims is necessary, and apportionment of the rent obligations among subtransferees raises extremely complex questions. Adoption of the rules of landlord and tenant law provides an answer—and we think a satisfactory one—to the problem of facilitating transfers of rights in the spectrum while protecting the government's interest in its purse.

Last, there is the problem of collecting rents when the original lessee transfers his whole property interest to another or divides his area or frequency band among others, saving no part of it for

61 It is said that the covenant to pay rent does not run to the subtenant because there is no privity of estate between the landlord and the subtenant. We can, we trust, adopt the results, without the reasoning, of the landlord-tenant cases.

himself. The easiest solution is to have the original lessee remain liable for the rent and depend on the transferees to actually pay it on pain of loss of their interest. However, it is also possible to impose liability on one of the transferees,as we propose to do in the case of excessive signal field strength.[62]

There are doubtless other general provisions that ought to be included in the organic law creating property rights in the spectrum that we have not discussed here. There must, for example, be provisions regulating the auction for the disposition of the TAS packages; authority must be delegated for the creation of the TAS combinations; and there must also be a detailed set of rules on the enforcement of the property rights created in the spectrum. The latter is dealt with in section B.

In general, however, we think the statute should leave many of the questions that will arise under the property system to resolution under state law. The primary functions of the proposed statute are to define property in the spectrum; to declare that the property interests are fully transferable pursuant to the purpose of creating a nationwide market in the spectrum; to set forth uniform enforcement procedures for protecting the property rights; and to encourage an economically efficient allocation of the resource by setting up an auction system for initial distribution of the TAS combinations. Most other questions of the operation of the property system can be left to state law.

Thus the statute should provide that state law shall control the transfer and enjoyment of rights in the spectrum, except as otherwise provided in the act. This provision is intended to make the statute preemptive both in its particulars and in the general scheme of creating a market in TAS combinations, but to leave to state law the innumerable questions that can arise over the use of the property. State law would control such matters as false and deceptive advertising and the effect of zoning restrictions on the location of transmitters. On the other hand, a state statute purporting to increase or decrease the permissible field strength of TAS owners would conflict with the federal statute and would be unenforceable. More subtle subversions of the federal scheme should also be declared ineffective, as, for example, a state statute seeking to interfere with market allocation of the spectrum by

62 *See* text accompanying notes 68–69 *infra*.

licensing users or by regulating the uses of certain frequencies. While some members of the scientific and economic community may question the ability of judges to administer such rules governing market processes, we believe these questions can safely be left to the courts. In fact little else can be done, for the mind of man cannot devise a statutory answer in advance for every question that will arise under the new property system.

B. Enforcement of Rights and Duties

1. Relationship between property-right definition and enforcement and exchange costs

The positive-limit definition specified the property right in the spectrum as the exclusive right to originate electromagnetic radiation twenty-four hours a day, within a designated geographical area, on a specified band, at a field strength not to exceed *X* v/m at or beyond the boundaries of the designated area and not to exceed *Y* v/m outside the specified band. This definition is preferred over the three alternative definitions suggested in chapter III, section C because it will reduce exchange and enforcement costs. In the following chapter our reasons for rejecting the alternative definitions are more fully explained. Although a discussion of out-of-band emission problems is omitted for simplicity, the legal considerations and conclusions applicable to interference in the area dimension are equally applicable to interference in the frequency dimension.

The residue definition. Under the residue definition each TAS owner is liable to the owners of the geographical areas on each side of him for all radiation crossing the respective boundaries in excess of *Z* v/m. This means that *A* is responsible for radiation originated by owners of geographical areas several territories removed from his own.

The principal objection to this method of defining the right is that enforcement costs are high relative to the positive-limit definition. Referring again to figure 5, suppose that the total field strength crossing the *BC* boundary from area *B* is 1.5 *Z* v/m, 0.5 *Z* v/m too high. This excess may be due to excessive field strength by *A* alone, by *B* alone, by *A* and *B* in combination, or by *A, B,* and other TAS owners to the west, northwest, and southwest. Since *C*'s property right is defined as the right against *B* not to have signals at field strength greater than *Z* v/m crossing the *BC* boundary, *C* must sue *B*, and *B* must then sue *A* and any other adjacent TAS owner at whose boundary the signal level is excessive. This multiplicity of litigation increases enforcement costs. It is no answer to say that all who contribute to the excess could be joined as defendants in one action, because the apportionment problem appears insoluble.

The zero-limit definition. The zero-limit definition of the property right would require that the cause of action for violation of the limit be based on a claim that the plaintiff had suffered some undue harm to his property rights from the defendant's operations. The analogy that comes to mind is that of nuisance, of which Prosser has said: "There is perhaps no more impenetrable jungle in the entire law.... It has meant all things to all men...."[63] While a lawyer is therefore chary about defining spectrum rights in a relational manner, some engineers have preferred the zero-limit definition because it avoids the problem of determining the value for the *X* v/m and *Y* v/m limits on signal field strength.

We concluded that the zero-limit definition would be unsatisfactory. The explanation of this conclusion is facilitated by the formulation of four alternative rules of law governing the rights and duties among TAS owners. Rules I and II are alternative specifications for the zero-limit definition. Rules III and IV are alternative specifications for the positive-limit definition. For reasons to be stated, rule IV is regarded as the preferred definition of the property right in spectrum use.

> Rule I: The signal strength originating in any TAS area shall not exceed zero v/m at any point on or beyond the boundaries of such TAS area. For violation of this duty, a TAS owner shall be liable for all pecuniary damages suffered by any other TAS owner as the result of such violation.
>
> Rule II: The signal strength originating in any TAS area shall not exceed zero v/m at any point on or beyond the boundaries of such TAS area. For violation of this duty, a TAS owner shall, upon proof by any other TAS owner of harm caused to him by such violation, be required to reduce his signal strength to a level that eliminates such harm to the complainant.

63 W. Prosser, Torts § 87, at 592 (3d ed. 1964).

> Rule III: The signal strength originating in any TAS area shall not exceed X v/m at any point on or beyond the boundaries of such TAS area. For violation of this duty, a TAS owner shall, upon proof by any other TAS owner of harm caused to him by such violation, be required to reduce his signal strength to a level not in excess of X v/m at all points on or beyond the boundary of such TAS area.
>
> Rule IV: The signal strength originating in any TAS area shall not exceed X v/m at any point on or beyond the boundaries of such TAS area. For violation of this duty, a TAS owner shall, upon proof of such violation by any other TAS owner, be required to reduce his signal strength to a level not in excess of X v/m at all points on or beyond the boundary of such TAS area.

To rules II, III, and IV the following may, and we think should, be added:

> In addition to the foregoing relief, complainant may recover pecuniary damages caused by such violation.

Rules I and II are unsatisfactory from any standpoint. Rule I will deny many users any remedy for interference because of the inability to prove pecuniary loss. The use of the spectrum is for communication; the monetary loss due to inability to communicate from time to time and at sundry places will often be difficult to quantify. How much did a taxicab company lose because one of its cabs could not hear the dispatcher for an hour, a day, or a week? What are the losses in revenues to an FM station that cannot reach a suburb in its TAS area from time to time? Rule II is probably worse, for extensive litigation will be required to develop the criteria for determining what the term "harm" means; moreover, every dispute between owners will have to be litigated to determine the application of the standard of harm to the facts of the case.

The obstacles to enforcement of either rule I or rule II inhibit transfers of rights. Until adjudication has established the precise limits of the rights and duties inhering in a TAS package, there will be nothing definite to buy or sell. Under rule I, the TAS owner suffering interference could pay the TAS owner causing the interference to reduce his field strength, but he will surely be tempted to try litigation before he tries bargaining, to avoid purchasing something he already owns. In the case of the TAS owner causing the interference, there will be little incentive either to buy the right to cause interference or to settle the damage claim arising from the interference until litigation has determined the pecuniary damages the complainant has suffered. The effect on bargaining of rule II is the same; the price at which exchanges take place will be influenced by the standard of harm that the courts evolve and by the application of the standard to each dispute between TAS owners. Bargaining will once again wait upon litigation.

These defects in rules I and II might suggest the development of an alternative rule to cover the zero-limit definition: The tribunal shall determine, as between competing TAS owners, which use is more valuable and fix field-strength levels in accordance with this determination. However, property rights for every user will be uncertain until an adjudication has determined them, and even then presumably res judicata would not apply to changed circumstances rendering a formerly less valuable use more valuable; enforcement costs will also be high because of the difficulty in judicially determining which use is more valuable. Moreover, the uncertainty about the property rights caused by the necessity for litigation will inhibit reallocation of the spectrum because there is nothing to sell prior to an adjudication; even thereafter the possibility of further litigation may make it difficult to agree on a price. Therefore, we concluded that property interests in the spectrum should not be assigned zero emission rights beyond the boundary.

Rules III and IV quantify the field strength of the signal beyond the boundary. It will be observed that under such a system, harm may or may not be made an element of the cause of action for violation of the duty not to exceed X v/m at or beyond the boundaries of the TAS area. Rule III incorporates the element of harm, and it does have some advantages. In the first place, harm is a familiar requirement of recovery in most areas of the law; for example, without proof of damage there can be no liability for negligence.

We have already pointed out, however, that the concept of harm in spectrum cases cannot be limited to proved pecuniary damages if relief is to be granted in numerous instances of interference. If harm were more broadly defined as interference with a TAS owner's communication with his receivers, it might be argued that harm, so defined as an element in the cause of action, promoted economic efficiency, since legal relief against field strengths in excess of X v/m would be granted only when the violation imposed an external cost on other TAS owners. When the excessive field strengths do not produce social costs greater than private costs, there is no distortion of resource allocation and no cause of action.

While rule IV lacks these advantages, the elimination of proof of harm gives it one significant advantage over rule III. Rule IV raises one, and only one, triable issue of fact: What was the signal strength at the time in question? While this is not always an easy question to answer (see subsection 5), it is simpler than the questions presented by rule III. Rule III not only inquires into the signal strength at the time in question but also asks what effect the excessive field strength had on the complainant's communications with his receivers.

Of course, under rules III and IV, just as under rules I and II, the initial assignment of rights may turn out to be suboptimal, in that another party may have a more valuable use for portions of the TAS package than the owner. The property system not only contemplates this circumstance, it expects it. The question is: Which of the four rules facilitates the reallocation of the right to the higher-valued use? We have already noticed that rules I and II strongly inhibit prompt and efficient reallocation. It seems to us that rule III suffers from the same defects as rules I and II. Even when defined as interference with a TAS owner's communication with his receiver, harm is a sufficiently vague concept that it can be given meaning not by statutory criteria but only after a long series of ad hoc adjudications. Even then almost every application of the doctrine of harmful interference will require an adjudication. It therefore appears that enforcement expense will be relatively high and that recontracting will be inhibited because of the uncertain outcome of litigation.

We would note further that rule IV is the rule applicable to trespass to real property. *A* may wish to erect a twenty-story office building on his land and encroach only a few inches on *B*'s geranium garden, but *B* has the absolute right, enforceable by injunction, to prevent *A*'s trespass. Despite this rule, the real estate market seems to work quite well and, where public necessity dictates, condemnation can be used to acquire *B*'s land. Of course, condemnation is also applicable to spectrum rights.

Rules for transactions. A system of property rights is not completely specified until the rights and duties concerning exchange of those rights have been clearly established. Many of these will be inherent in the general legal framework, others will follow directly from individual rights and duties in the absence of exchanges, and some must be promulgated specifically for these transactions.

We noted in chapter II that a rights definition under which transactions between two parties can harm third parties is usually unsatisfactory even when a rule requiring consent and compensation of these parties is included. With or without such a rule, the exchange and enforcement costs required to reach economic efficiency will be substantially the same. What is needed is a definition that reduces the potential for harm to third parties, thus reducing the need for multisided transactions. No single definition will necessarily handle all types of exchanges most economically. The major varieties of exchanges are considered below with regard to flexibility and third-party protection.

(a) *Changing field-strength limits.* Referring back to figure 1, suppose that for the TAS areas centered on San Diego, Los Angeles, and Santa Barbara, all sharing the same band of frequencies, the field-strength limits are *X* v/m for all points outside each area. If these limits are being enforced, the San Diego owner's signal will be no more than *X* v/m at the San Diego–Los Angeles boundary, and, under normal conditions, considerably less than *X* v/m by the time it arrives in the Santa Barbara area. Now suppose that the San Diego owner buys from the Los Angeles owner the right to have signals up to 25 *X* v/m in the Los Angeles area. Because the San Diego owner is no longer confined to the *X* limit at his Los Angeles boundary, he may actually achieve the limit of *X* at the Santa Barbara boundary, but he cannot exceed this without the permission of the Santa Barbara owner, even if this means that his signal can only be 15 *X* v/m at the San Diego–Los Angeles boundary. A TAS owner—for example, the San Diego owner—desiring to increase his power or otherwise modify his own operation is required to negotiate with all parties whose interests will be substantially affected, and with no others—that is, with those in whose TAS area his signal strength will exceed *X* v/m.

Compare the above with a similar situation under the boundary–positive-limit definition where field strength is limited to *X* v/m only at the boundaries of an area rather than at all points outside the area. A sufficiently large change in the limit at the Los Angeles–San Diego boundary will greatly increase the field strength that the San Diego owner achieves in areas far removed from the Los Angeles–San Diego boundary. The definition must specify who must agree to such a change—that is, from whom the right shall be bought. If only the Los Angeles owner's consent is required, there is substantial potential for harm to the owners of areas in Santa Barbara and beyond. If the consent of specified third parties is required (for example, the owners of all adjacent and next-adjacent areas), exchange costs are raised significantly.

Under the first rule, the value of each TAS owner's combination is subject to substantial uncertainty because exchanges among other TAS owners may alter its physical properties. This "technological externality" can, of course, be prevented by the affected TAS owners' banding together to "buy off" the Los Angeles and San Diego owners, but the exchange costs of doing so will be substantial and may well exceed the reduction in the value of the affected packages.

The alternative of requiring the consent of specified third parties prior to any transaction reduces the uncertainty about the value of the TAS package, but significantly raises exchange costs. Exchanges will be postponed or forgone unless the gain from the reallocation of resources is greater than these costs.

The problem with the boundary-positive-limit definition is that there is only a single limit on field strength—that is, X v/m *at the boundary*; all owners rely on this for preservation of the physical properties of their TAS packages and all can be affected by a change in the limit. The positive-limit definition provides a separate limit in each area which protects the owner of that area; it brings into each transaction exactly and only those who are significantly affected, while the boundary-positive-limit definition, either with or without some rule requiring the participation of third parties, seems to involve substantially higher exchange costs to achieve the same efficient allocation of resources. Therefore, the boundary-positive-limit definition is eliminated from further consideration.

(b) *Combining areas and changing inputs.* Another major type of transaction is the combination of two or more TAS areas. Suppose the San Diego owner purchases the Los Angeles TAS package for the same band. He can then shut down the Los Angeles transmitter, increase the power of his San Diego transmitter so as to achieve as much as X at the Santa Barbara boundary, and thereby benefit from both the increased power and the elimination of interference in his band from the Los Angeles area. Alternatively, he could use both transmitters for the same signal, again with increased power for the San Diego transmitter. However, the Santa Barbara owner is still allowed to exclude signals greater than *X* v/m from the old Los Angeles area and from the old San Diego area; common ownership does not permit the San Diego owner to add the field-strength limits and apply them to a single transmitter.

Although not strictly a transaction, the unilateral alteration of antenna height, location, or power has some of the same potential for harm to nonparticipating parties, because these changes can affect the rate at which a signal's field strength attenuates while crossing other TAS areas. As long as these changes do not result in a signal that exceeds *X* v/m outside the originating area, the other TAS owners have no cause of action even though they may incur some cost in the form of increased interference. However, an appropriately determined field-strength limit and the incentives favoring low antennas noted in chapter III should keep these costs small.

2. Breach of duty and standard of care

A wide variety of standards for measuring breach of duty is encountered in the law. Criminal liability usually depends on the forbidden act being done intentionally. A landowner owes a duty to a trespasser only to avoid infliction of wanton or willful injury. Negligence is defined as the failure to exercise ordinary care—that degree of care that an ordinary, prudent person would have exercised under the same or similar circumstances. In personal injury cases courts are beginning to resort to the standard of strict liability on the ground that this standard results in more efficient resource allocation by requiring an entrepreneur to bear the full social cost of his enterprise.[64] For this reason and others, we believe that the standard of strict liability should be applied to breach of duty by a TAS owner. Moreover, the adoption of a simple rule will reduce enforcement costs and facilitate exchanges.

Under a strict liability standard the only issue of fact in litigation over TAS rights will be whether or not the radiation

64 *See, e.g.,* Seely v. White Motor Co., 63 Cal. 2d 9, 18, 403 P.2d 145, 151, 45 Cal. Rptr. 17, 23 (1965); Greenman v. Yuba Power Prods., Inc., 59 Cal. 2d 57, 377 P.2d 897, 27 Cal. Rptr. 697 (1963).

originated by the defendant exceeded *X* v/m at or beyond the defendant's boundary at some given time. Any less severe standard will require the tribunal to determine such questions as (1) did the defendant intend to exceed the maximum field strength; or (2) would an ordinary, prudent person in the circumstances of defendant know, or have reason to know, that his signal strength would exceed the maximum; or (3) did the defendant believe in good faith that he was operating within the permissible signal strength? In addition to increasing litigation costs, standards such as these make the property right less certain and to that degree interfere with the exchange process. An analogy to land law is again applicable: If *A* puts his fence on *B*'s land, it makes no difference whether *A* acted intentionally or by mistake, carefully or carelessly, in good faith or bad; the fence must come down if that is *B*'s wish. The price of obtaining the right to maintain the fence does not depend on a subjective or objective assessment of *A*'s state of mind.

It has been argued that under some circumstances the strict liability standard might result in economic inefficiency. The argument starts with the assumption that *A*'s radiation in excess of the field-strength limit interferes with the communications of a large number of other TAS owners. It further assumes that it would be cheaper for the interferees to adjust their systems—for example, by raising their antennas or increasing their transmitter power—than it would be for *A* to make the necessary modification in his system, but that because of the large number of interferees *A* will be unable to pay them to make the adjustment. Therefore, when one interferee sues *A*, the strict liability standards will require *A* to adjust his own transmitter, with resultant economic loss. The argument concludes that the preferable solution is a less severe standard of care, such as negligence, which would permit the court to hold *A* liable only if it is cheaper for him to adjust his equipment than for all of the interferees to adjust theirs.

We are not persuaded by this argument. First, it must be noted that calling the standard one of negligence is putting new wine into an old bottle. The standard suggested is not that of a reasonable man, but that of an economic man possessed of all the facts. That alone is not a criticism, but recognition of it may avoid confusion. Second, for the argument to succeed, it must be assumed that the aggregate costs of equipment adjustment for all the interferees are less than those for *A*, the interferer. Even though the particular lawsuit may involve only *A* and one interferee, ***B***, and even though it may be cheaper for ***B*** to modify his equipment than for *A* to do so, there is no loss of efficiency in making *A* bear the costs of adjustment so long as the aggregate costs of adjustments by all interferees are greater than *A*'s adjustment costs. Since the argument posits so many interferees that exchange costs prevent *A* from buying the right to exceed the field-strength limit, we are doubtful that the aggregate adjustment costs of all these interferees will often be less than the adjustment costs of *A* alone.

Third, as the number of interferees is reduced to make the argument more plausible, their aggregate adjustment costs are also reduced and could conceivably become less than *A*'s adjustment costs. At the same time, however, exchange costs are also reduced, making it increasingly possible for *A* to buy the right to radiate at increased field-strength levels. In short, it appears likely to us that whenever the aggregate adjustment costs of the interferees are less than *A*'s adjustment costs, the number of interferees will be sufficiently small that exchanges can be effected by *A* to achieve the efficient solution. A final defect of the argument is that it ignores the cost of determining whose adjustment costs are lower, *A*'s or the interferees'. It is clear that this determination cannot be obtained at zero cost.[65]

3. Who can enforce?

In considering this question, two categories of illegal radiation must be distinguished: radiation by unauthorized emitters, and radiation in violation of duty by TAS owners.

As for the first category, the statutory scheme makes the radiation rights of the TAS owners exclusive. Therefore, any unauthorized operator radiating within an assigned TAS package is guilty of a breach of duty and can be shut down. The practical effect of this rule is to require miscellaneous emitters—such as operators of diathermy machines and arc welders—to shield their equipment, shut down, or buy TAS rights.[66] We propose to allow any TAS owner or assignee affected by such unauthorized emissions to sue to enjoin the unauthorized radiation.

65 A similar problem of information costs arose in the optimization of the initial TAS packages. *See* text accompanying note 40 *supra*. *See also* page 1536 *supra*.

66 Alternatively, the police power could be used to eliminate such violations. *See* note 25 *supra*.

As for the rights *inter sese* of TAS owners, we have concluded that only persons holding TAS rights should have the power to enforce the duty not to exceed the prescribed field strength. This rule denies rights to owners of receivers. Thus if *A*, by exceeding *X* v/m at his boundary, makes it difficult for owners of receivers in ***B***'s territory to receive ***B***'s signal, ***B*** and only ***B*** has the right to require *A* to reduce his signal strength. Allowing owners of receivers to enforce the duty would seriously interfere with market transfers between TAS owners because of the high costs of transacting with large numbers.

For most firms owning TAS rights these rules should be unobjectionable. Land-mobile and other point-to-point communications operators own both the transmitter and the receivers. Difficulties may arise in radio and television broadcasting, where the station does not obtain revenues from receiver owners. Although the business of the broadcaster depends upon the audience he reaches, the value of marginal viewers to advertisers, and hence to the station, may be less than the viewers themselves would be willing to pay in enforcement costs to receive the broadcasts. Despite this shortcoming, which also exists in the present system in that advertisers are proxies for listeners, the desirability of avoiding high exchange and enforcement costs seems to justify the denial of rights to receiver owners.

Several analogies support the denial of enforcement rights to a large class where there exists a smaller group with the incentive to enforce the right for the benefit of all. For example, the right of a shareholder in a corporation to enforce the corporation's causes of action is severely limited. The customer of a municipal corporation that supplies water cannot enforce the corporation's water rights. The doctrine of standing to sue limits the right of suit by those adversely affected by the wrongdoing of others to a small, representative group whose interest is more directly affected and whose position enables them to represent others. We believe that radio and television transmitter owners will have a sufficiently strong interest in enforcing their spectrum rights that it is unnecessary to allow the larger class of owners of receivers to sue.

4. *Who is liable?*

The question of enforcement of rights and duties becomes difficult only after assignments of TAS packages have been made.

Prior to assignment the problem is simple: Each TAS owner can enforce his right to be free of excessive radiation against any other TAS owner guilty of a violation of the signal-strength limits. Where the area component or the spectrum component has been subdivided or reconstituted, the problem of assigning liability for excessive signal strength becomes more complicated. Such changes of ownership do not affect the basic protections provided by the property system; the time-area-spectrum specifications in the certificate of ownership still provide the basis for determining whether or not a violation has occurred. In the discussion that follows we divide the problem into two parts: assignments that maintain the original TAS package, and those that eliminate the original package and create new TAS packages. Within the area or spectrum component that has been divided, the contract between assignor and assignee will be the source of protection. Our concern here is with the third-party effects of bilateral agreements.

Assignments retaining the original TAS package. Suppose *A* owns 200 kHz of bandwidth at 88.0 to 88.2 MHz in the Los Angeles TAS area. He sells the south half of the TAS area to *Q*. *A* and *Q* radiate at power levels that produce a field strength of 2 *X* v/m in *B*'s TAS area to the east. What are the liabilities of *A* and *Q*?

The law could require *A* and *Q* to reduce their power levels so that total field strength did not exceed *X*, the reduction being apportioned between them on the basis of the ratio that the size of their subdivisions bears to the size of the Los Angeles TAS area. Thus, if *A* and *Q* each owned half of the Los Angeles area, *A* and *Q* would each be required to reduce their power equally, to produce not more than *X* v/m at or beyond the boundary. Undoubtedly other ways of apportioning the required reduction could also be found.

Rather than apportioning between *A* and *Q*, the law could place the entire burden of a reduction on one of the two parties. The controlling principle could be either "first in time is first in right" (thus making *Q* bear the burden) or "last in time is first in right" (leaving *A* with the burden). We believe the better rule is to allocate the responsibility entirely to one party or the other. Apportionment is likely to promote litigation and increase exchange costs. Placing the full burden on *A* or *Q* avoids this, and has no countervailing cost. If the first user, A, bears the burden he will make provision for that burden in the assignment contract with Q. He may shift the risk entirely to Q; he may specify the rights that Q obtains in such a manner that excessive signal strength cannot be legally caused by Q; or he may retain the burden. If a rule is adopted that places the burden on Q, contracting will again ultimately allocate the risk. It therefore seems to make no difference where the burden is initially placed, so long as it is placed on one party or the other and is not apportioned.

Subdivisions and recombinations of TAS packages by market processes will probably be more complicated than the simple north-south division of the Los Angeles TAS area example just given. However, the priority system still appears to be satisfactory for protecting adjacent TAS owners. For example, no matter how many subareas the Los Angeles TAS area may be divided into, adjacent area owners will be protected against radiation in excess of X v/m at or beyond the original boundaries of the Los Angeles TAS area, and they will have a cause of action against operators in the area in accordance with the priority rule adopted. Thus, if the last-in-time-is-first-in-right rule is adopted, and the Los Angeles TAS area has been divided into twelve subdivisions owned by A_1 (the original TAS owner) through A_{12}, a complaining adjacent TAS owner, B, will begin with A_1, and join in the lawsuit to enforce his right to as many assignees as necessary to reduce the field strength originated in the Los Angeles area to X v/m. Though the liability for excess field strength is determined by a priority system, this does not mean that B must sue each assignee successively, beginning with A_1 and ending eleven lawsuits later with A_{12}. Since the remedy for breach of duty extends to shutting down the offender's transmitter if necessary, B may join in one suit A_1 and that number of his assignees (in chronological order) whose aggregate signal strength, if eliminated, would reduce total radiation from the Los Angeles area to X v/m.

Moreover, an expansion of the third-party-beneficiary doctrine would further reduce litigation costs. Suppose the San Diego area were subdivided into four parts with C_1 (the original owner), C_2, C_3, and C_4 as the respective owners. Each contract specifies that the subdivision owner will limit signal strength to 0.5 X v/m at the original TAS boundary or beyond. If C_4's signal has a field strength of 0.75 X v/m at the boundary, raising the aggregate field strength of the signal to approximately 1.15 X v/m,[67] B should be able to sue C_4 directly as a third-party beneficiary of the contract obligation undertaken by C_4. In this case, although B could enjoin C_1's radiation, the ultimate liability is C_4's, and both B and C_1 should be able to sue C_4 in a single lawsuit to correct the infraction.

Returning to the hypothetical division of the Los Angeles area into twelve parts, suppose that the original TAS owner, A_1, sells his remaining subdivision to A_{13}; suppose further that A_{13} violates the agreements A_1 had with A_2 through A_{12} regarding A_1's field-strength levels. It might seem in this case that the inverse order of priority rule would operate to benefit the wrongdoer. But a simple solution to this problem is to treat A_{13} as taking his TAS combination subject to the contract burdens imposed upon it by his predecessors in title. Thus if B sued A_2, A_2 could enforce against A_{13} the contract duties that A_1 undertook in all of the contracts of assignment, with A_2 through A_{12} as well as the duties A_{13} personally undertook in his contract with A_1. This extension of the third-party-beneficiary doctrine would also permit B to sue A_{13} directly to enforce the contract duties A_1 assumed with A_2 through A_{12} as well as the obligations that A_{13} undertook in his contract with A_1. Thus even in the case of multiple subdivisions of TAS areas a strict priority rule coupled with an expanded third-party-beneficiary doctrine will serve to allocate liability efficiently and will therefore be preferable to a rule of apportionment of liability among assignees.

We are now in a position to deal briefly with rights of assignees among themselves. These rights will depend strictly on contract, since by hypothesis we are dealing with the owners of parts of an original TAS package. Nevertheless, the expanded third-party-beneficiary doctrine is also helpful here. Any assignee whose contract protects him from radiation within his subdivision in excess of a specified field strength should have the right to proceed not only against his assignor-promisor, but also against any other assignee whose violation of his own contract contributes to the excessive field strength. Also, any assignee of a portion of the TAS package who is subjected to excessive field strength from an

67 The resultant field strength X_{total} due to constituent field strengths X_1, X_2, X_3, and X_4 is given by the formula $X_{total} = \sqrt{X_1^2 + X_2^2 + X_3^2 + X_4^2}$.

adjacent TAS area will have a cause of action. Thus if the Los Angeles TAS area has been divided into twelve subdivisions and the adjacent San Diego area has been divided into four areas owned by C_1 through C_4 respectively, C_1 through C_4 have exactly the same rights as C_1 initially had for legal relief against excessive field strength originating from the Los Angeles area. Obviously, C_1 through C_4 can join as plaintiffs in a single action if they wish.

Assignments creating new TAS packages. Thus far we have been concerned with subdivisions for which the original TAS packages continue to provide the basis for determining whether a violation has occurred. This necessity to refer all subsequent combinations of rights back to the original package is likely to become increasingly cumbersome and might eventually be a serious restraint on transferability. For this reason, it would be desirable to have a form of reconstitution in which distinct, independent, fully specified TAS packages, each with its separate rights and duties, could be created from the original package, which would then cease to exist. These new packages would provide the basis for determining whether a violation had occurred. It will be economically efficient to permit this reconstitution if rules can be found that provide third parties with approximately the same protection they had under the original packages without requiring their participation at every step, and that do not contain so many restrictions that the newly created parts are less valuable than the previous whole. Rules that appear to meet these qualifications are discussed below for subdivision of the area and spectrum dimensions; subdivision of the time dimension does not involve any apparent difficulties.

Suppose the Los Angeles area (for a particular time and spectrum) is reconstituted into subsequently independent areas centered on Long Beach and Pasadena. The field-strength limit in Santa Barbara, San Diego, and other areas on signals from the Los Angeles area must be reconstituted as well. The simplest way to accomplish this is to give the same fraction of the former limit to a given subdivision at every point outside the area. For example, the Pasadena area could be assigned a limit of 0.6 X at every point in San Diego, Santa Barbara, and other areas, with the remainder assigned to the Long Beach area.[68] This would in effect

68 Because field strengths "add" according to the square root of the sum of their squares, the remainder is .8 X, not .4 X: $\sqrt{(.6\,X)^2 + (.8\,X)^2} = X$.

create two new areas whose net effect in other areas could never exceed that of the original.

Owners of other areas will usually be better off than before under this rule. The transmitters in the two new areas are likely to be geographically separated; this means that in places where one signal is relatively strong the other will be relatively weak. Since each signal separately must meet its limit at its strongest point outside the area, their total field strength will usually be less than the previous limit. This provides an incentive not to reconstitute since the same two transmitters under a single operator would be measured together rather than separately, and could take advantage of this counterbalancing high-low effect to operate at somewhat higher power.[69] Therefore, the following rule is suggested: Any more detailed subdivision of the field-strength limits is allowed provided that the total permitted field strength from the new areas does not exceed at any point in any other area the field strength previously permitted from the old area. This would permit the Pasadena area to have a larger fraction of X in areas close to it and a smaller fraction in areas close to Long Beach.

A similar analysis can be made of reconstitution of the spectrum dimension of the TAS combination. Suppose the San Diego owner reconstituted his original unit of spectrum, say 25 kHz, into two channels each covering the entire area. The field-strength limit in other areas must also be divided. The field strength X is an integral over the 25-kHz band; therefore, the sum of the integrals over the two 12.5-kHz bands must not be more than X. An equal division would assign to each of the new channels a limit of 0.5 X over its band. Should the total in Los Angeles exceed X when integrated over 25 kHz, the Los Angeles owner would have to look at the two new 12.5-kHz channels to determine which had exceeded the 0.5 X limit. This relatively simple rule permits the net effect of the two independent channels to be no different than two identical channels under a single operator.

5. *Technical aspects of measuring violations*

Methods currently used for FCC measurements of field strengths of AM standard-broadcast stations[70] and for locating unlicensed transmitters should also serve to detect and prove violations of

69 This problem is one that the residue definition handles much more neatly. Rights in that system are defined in terms of *total* radiation at a given point on the

TAS rights; it would only be necessary to supplement these methods by regulations specifying uniform techniques and standards of measurement. The purpose is not to predict the field strength of a signal, but only to answer the easier question: Does the signal exceed the limiting value?

Identification of interference sources. Assume that a dispute requiring a field-strength measurement arises because of the emergence of a troublesome level of interference. The first step is to identify the source of interference. The most obvious means of identification would be through the recognition of call signs, unique message content or format, or type of modulation. In many cases such data are readily available through the signals causing the interference. Public records showing TAS ownership and assignments will provide additional information; of particular interest will be recontracting activities that have resulted in new transmitters in neighboring areas or bands. With this information direct communication or negotiation between the offending and offended parties may solve the problem.

In cases where the interference is intermittent, a receiver with a voice-actuated recorder is capable of monitoring while unattended over extended periods of time. A more sophisticated device will record the field strength of the signal simultaneously with the messages. An alternative technique is the use of a directional antenna to identify the location of the interfering transmitter. Care will be required in making directional measurements because of the prevalence of multipath propagation in the frequency range considered in this essay, but in most cases not more than ten measurements should be required.

Measurement of interference. After the source of the interference has been located the offended party can determine whether the interfering signal is authorized to transmit on his band of frequencies or channel in another area (co-channel interference) or on an adjacent channel in his area (adjacent-channel interference). The offended party would then have to conduct a survey of signal field strengths in his area and frequency in order to prove that the interference had exceeded the *X* v/m limit on co-channel interfer-

border; this automatically takes account of the counterbalancing of near and distant signals. Division of responsibility is accomplished by each new area simply assuming responsibility for all radiation crossing its borders into other areas.

[70] *See, e.g.,* FCC Rules & Regulations § 73.186, 47 C.F.R. § 73.186 (1968).

ence or the Y v/m limit on adjacent-channel interference. Such a field survey would require two to four man-days for data collection and about an equal amount of time for data analysis. Repeated measurements could take advantage of previously collected data and would require a less thorough survey, although the identification of the interference source would have to be verified to assure that new sources of interference had not appeared.

Although not directly related to enforcement, we note that there will be a need for similar field surveys before a purchase is made of all or part of a TAS package. The prudent purchaser would obtain such a survey to assure himself that the TAS package was as represented. He might also wish to transmit a test signal during the survey to assure that the service contemplated could be provided by the TAS package in question. It is likely that independent consulting firms specializing in field surveys will be formed to provide such services to TAS owners and purchasers.

6. Enforcing tribunal

The choice of an enforcement agency is between a federal regulatory agency such as the FCC and the courts. We believe the courts, both state and federal, should be given the responsibility. Courts are available locally all over the country; using them will save travel expenses. Moreover, the courts will be able to act more expeditiously than will a single agency responsible for all interference litigation. There is court congestion in metropolitan areas, but if the recommendations that breach of duty should be based on strict liability and that the proper remedy is an injuction are accepted, the issues to be tried will be simple, pretrial discovery should not be time consuming, and the case can be tried by a judge on the equity side of the court. Some suits for excessive radiation will undoubtedly seek damages as well as an injunction, and the courts have had more experience in handling damage cases than has the FCC or than a new agency would have. The courts may not have as much expertise in electromagnetic engineering as the FCC, but courts reguarly try technical issues and seem to do a competent job.[71]

[71] Even if an administrative agency were utilized, the courts would become involved either by a proceeding to enforce the order of the agency or by a proceeding to

7. Remedies

We propose that an injunction be granted as a matter of right to any complainant who establishes that his rights have been violated by another TAS owner or assignee or by an unauthorized broadcaster. For the same reasons that we rejected harm as an element in the cause of action for excessive field strength, we would abolish the traditional equity requirement that the remedy at law must be inadequate before an injuction will issue. Exchanges will be facilitated and enforcement costs reduced if the nebulous issue of damages is not made a condition precedent to relief. Of course, where a party is prepared to pay the extra cost of litigating this issue, he should be allowed to recover any pecuniary damages he has suffered, thereby forcing the wrongdoer to pay the social costs of his action.

We have rejected criminal penalties and fines as additional sanctions because they seem inappropriate to the offense committed. Injunctions are the standard means of enforcing many rights in real property; for example, courts will enjoin recurring trespasses, excessive use of easements, and nuisance interference with the enjoyment of land. The harm caused by excessive radiation likewise constitutes interference with another's use of his property; Thus the appropriate remedy is to enjoin the interference in accordance with the property rights created by the proposed system.

Two concerns about the use of injunctions are the requirement that the violation of an injunction be willful before a court will put the defendant in jail for contempt, and the danger that the courts will frame the injunction in terms of inputs such as antenna height and transmitter power rather than outputs (field-strength levels). We believe both of these matters may be entrusted to the courts with some confidence in their ability to execute the purpose of the statutory scheme, but the statute should contain a provision on remedies along the lines of the following model, and the legis-

review the agency's action. Generally, Congress has required that an administrative agency obtain an enforcement order from the courts before the offending party can be punished for failure to comply, *see, e.g.,* 29 U.S.C. § 160(e) (1964) (NLRB), or has allowed the offending party to obtain judicial review, *see. e.g.,* 15 U.S.C. § 21(c) (1964) (FTC). Alternatively, the courts would have to review the agency action in any suit brought to punish the offending party for violating the order of the agency.

lative history should show the purpose of the provision as the subsequent comments indicate:

> Section ——. *Enforcement.* Wrongful operation is defined (1) as the emission by a TAS owner of electromagnetic energy in violation of this Act or of the terms of the TAS owner's certificate of ownership and (2) as the emission of any electromagnetic energy by any person not authorized to do so under this Act and not exempt hereunder. Upon proof by a TAS owner, or assignee, affected thereby of wrongful operation the court shall require, by injunction or other equitable means calculated to prevent the wrongful operation, the person responsible for the wrongful operation to cease such operation. Offending TAS owners shall be required to operate in compliance with the terms of this Act and their certificates of ownership; unauthorized operators shall be required to cease emitting unauthorized electromagnetic energy.
>
> *Comments:*
>
> 1. Upon proof by a TAS owner that an unauthorized operator is causing emissions to be received within his territory and band the complainant shall be entitled to a permanent injunction against the unauthorized operator requiring him to cease operation entirely. [Another section of the statute could provide that the complainant shall also be entitled to whatever damages he can prove.]
>
> 2. Upon proof by one TAS owner that another TAS owner is violating the statute or the terms of his certificate of ownership and thereby causing unauthorized emissions to be received within the TAS area and band assigned to the complainant the complainant shall be entitled to an injunction requiring the offender to comply with the statute and the terms of his certificate. In view of the fact that the purpose of this statute is to encourage the economically efficient use of resources devoted to communication by electromagnetic energy, the preferred form of injunction is an order requiring the defendant to operate his equipment in compliance with the terms of his certificate of ownership so as not to violate the rights of the complainant.
>
> a. *Example: A* has the right to originate signals within his TAS area at a field strength not to exceed *X* v/m at or beyond the boundaries of such area. *A* is operating his equipment so that a signal of 2 *X* v/m is received in *B*'s TAS area. *B* is entitled to an injunction requiring *A* to modify his manner of operation so as to reduce the field strength of his signal at or within *B*'s boundary to a maximum of *X* v/m. While the court could accomplish this objective by prescribing the antenna height and transmitter power used by *A*, injunctions in such form should be avoided because they are inflexible and may result in an uneconomic use of resources.
>
> b. It is recognized that an injunction in the form recommended is less precise than injunctive decrees ordinarily are. For this rea-

son, it is desirable to allow the offender a reasonable amount of time for experimentation in complying with the decree. If, after the expiration of the reasonable time period set forth in the decree, the offending TAS owner has still not complied with the injunction, the remedy of contempt is available for dealing with willful violations of the order. In other cases it would be appropriate for the court to appoint a custodian of the offender's transmitting equipment, with directions to modify the operation in the most economical way to achieve compliance with the court's decree. This is provided for in the section above by the language "or other equitable means calculated to prevent the wrongful operation. . . ." Costs for measuring signal strength to determine compliance with a decree should be assessed against the offender.

8. Prescription

A further question that must be considered is the applicability of the doctrine of prescription to TAS rights. If prescription were to apply, persistent violation by one TAS owner of the rights of another, where redress had not been sought by the injured owner, would result in a transfer of a portion of the TAS rights of the injured owner to the offender by operation of law. To illustrate, suppose TAS owner *A* radiated at a field strength of 2 *X* v/m in the TAS area owned by *B*. If after expiration of a statutory period —either fixed by federal law or determined under state law—*B* had not vindicated his right in court, *A* would have a prescriptive right to continue his radiation at the higher level.

Despite our earlier recommendation that state law should generally be applied to resolve private disputes over spectrum use, we have concluded that the organic statute should forbid the acquisition of prescriptive rights in the spectrum for three reasons. First, signal variability creates problems. Excessive radiation is unlike adverse possession, where *A* locates his fence 25 yards into *B*'s property. A typical spectrum case might involve evidence showing that *A* transmitted a signal at 200 percent of *X* v/m 10 percent of the time, at 150 percent of *X* v/m 30 percent of the time, and at *X* v/m 60 percent of the time. How are *A*'s prescriptive rights quantified? The costs of determining the answer to this question are likely to be quite high. Second, recognition of prescriptive rights might necessitate boundaries that would disrupt the geographical scheme of the statute. Thus if *A* is transmitting a signal of 2 *X* v/m, he has created a new field-strength contour inconsistent with his old boundary. Is this contour to be recognized as the new boundary of *A*'s designated geographical territory? If so, the boundary will be irregular and not clearly delineated. Although we could allow the creation of prescriptive rights in field-strength levels and not formally change the geographical boundaries, the practical effect of altering field-strength levels is to change boundary lines as well.

Third, recognizing prescriptive rights would have an adverse impact on transfers of TAS packages. Exchange costs are lower when property rights are fully defined and on record, allowing the purchaser to know exactly what he is getting. Bargaining becomes more difficult when there are valid legal rights not reflected in the registry. The complete elimination of prescriptive rights removes this obstacle to the operation of the market. If, however, prescriptive rights are to be recognized the statute should provide that no prescriptive right not reduced to judgment and recorded in one central registry shall be binding on transferees of the "servient" interest—that is, the interest that is subject to the prescriptive right. This suggestion goes further than merely protecting the bona fide purchaser of the servient estate; it protects all TAS transferees. The reason for this is that the determination of what the purchaser knew or ought to have known will involve expensive litigation because of the variable nature of electromagnetic wave transmission. For example, the issues in the litigation might include: How many measuring devices should the transferee have used? How many locations should he have tested? How long should he have tested? To avoid the costs of answering these questions and to avoid the damper on transferability that will be caused by the existence of these questions, we think that the records kept in the central registry should be the sole measure of existing TAS rights.

The proposition that prescriptive rights in the spectrum be denied recognition is supported by analogy to the water law of several Western states. Many appropriation states moved from the self-help creation of water rights by diversion and beneficial use to an exclusive permit system, whereby water rights are created only pursuant to a permit issued by an administrative agency. One compelling reason for the change was to provide a system under which all rights to use water were on record. Thus the laws of several states hold that prescriptive rights to water cannot be obtained under the permit system.[72]

The above discussion is not intended to suggest that the statute of limitations never applies to TAS rights. Certainly stale damage claims should be barred after the lapse of a period of time; this we can safely leave to state law. All that is proposed here is that the statute creating the property system specifically bar the acquisition of prescriptive rights in the spectrum.[73]

72 *See, e.g.*, KAN. STAT. ANN. § 82a–705 (1964); NEV. REV. STAT. § 533.060 (1967); UTAH CODE ANN. § 73–3–1 (1961).

73 Although stale damage claims can be barred, the courts could not use laches to bar a suit for an injunction, even though the injunction is an equitable remedy. To allow the courts to use the laches doctrine in this manner would simply create prescriptive rights by another route.

V. Disposition of Government-Controlled Resources: A Historical Summary

We have advocated here a change in the management of the electromagnetic spectrum from a government-administered system of allocation to a market system. We have further attempted to formulate a system of property rights in the spectrum that will permit market exchanges to take place. Two other tasks remain: first, to demonstrate the historical basis for our proposal that the government should relinquish the spectrum to private ownership; and second, to show how the transition from the administered system to the market system can be arranged.

A. Historical Patterns of Resource Allocation

1. Disposition of federal land

The major portion of all real property ever owned by the federal government has now been disposed of in fee simple absolute; the largest portion was given away. Sixty-eight million acres were donated as military bounties;[74] other donations made to encourage specific activities thought to promote the development of the nation include wagon-road grants,[75] railroad grants,[76] grants for reclaiming swamp and overflowed lands,[77] and educational grants.[78] The largest number of dispositions of the public domain

74 *See* B. HIBBARD, A HISTORY OF THE PUBLIC LAND POLICIES 132 (1939).

75 *See, e.g.,* Act of Feb. 28, 1823, ch. 16, 3 Stat. 727.

76 The statutes are too numerous to cite. A total of 129 million acres was disposed of for this purpose. B. HIBBARD, *supra* note 74, at 264.

77 *See, e.g.,* Act of Sept. 28, 1850, ch. 84, 9 Stat. 519. Over 63 million acres were transferred by this act, an area greater than the combined areas of New York, New Jersey, New Hampshire, Massachusetts, Vermont, Connecticut, and Rhode Island. B. HIBBARD, *supra* note 74, at 275, 287.

78 *See, e.g.,* Act of July 2, 1862, ch. 130, 12 Stat. 503 (the Morrill Act).

were made under the Homestead Act of 1862[79] and the Desert Land Act,[80] which granted land titles, on the payment of nominal fees, to any farmer who settled on the land and cultivated it for a period of three to five years. Over 213 million acres were transferred from government to private ownership under these two statutes alone.[81]

Of importance here is the nature of the property interest received by the grantee under these acts. When the transfers from government to private ownership were made, there existed a ready-made body of property law defining the rights and duties of the new owners. This was, of course, the English land law from which had evolved the concept of the fee simple absolute. In accordance with this concept the landowner had four fundamental rights he could exercise with respect to his land: the right of use, the right of transfer, the right to be protected from invasion by others, and the right to time-tenure in perpetuity. In addition, the property interest was defined on the ground, so that the rights extended to fixed spatial boundaries.

As we have pointed out before, these rights made it possible for the use of the land to be reallocated through the market. Thus a parcel of 160 acres acquired as a farm under the Homestead Act in the last century may now be subject to a variety of uses, many of them undreamed of when the original patent was issued. The land might now be the site of a shopping center owned by an investment syndicate and rented under percentage leases to a variety of businesses, a condominium in which each apartment owner holds title to a cube of airspace with cross easements in public areas and subject to an operating agreement, or a residential subdivision with easements for services such as water, electricity, sewage, and gas, and with private restrictions controlling setback lines, minimum living space, architectural design, and use and occupancy. The point is that, by disposing of its real property in fee simple absolute, the government surrendered its powers to allocate the use of the resource—short of the use of eminent domain—giving way to private decision making in response to market forces.

This is not to say that private decision making has been free of

79 Ch. 75, 12 Stat. 392.

80 Act of Mar. 3, 1877, ch. 107, 19 Stat. 377.

81 B. Hibbard, *supra* note 74, at 398.

government regulation or that it should be. Land use has been and is still the subject of a wide range of governmental restrictions, most of which have been imposed under state, and not federal, authority. These restrictions have been adopted to achieve a variety of objectives, including the efficient allocation of resources. The latter objective has been a major source of land-use restrictions. Thus subdivision and zoning controls may be used to prevent the imposition of external costs on others. The Rule Against Perpetuities prevents the removal of land from commerce for an excessive period of time. The law of nuisance and pollution-control laws may save high exchange costs by making it unnecessary for adversely affected parties to join together to buy off the offender. Controls on oil and gas production seek to solve the "common pool" problem.

Such limitations on the rights of landowners were, significantly, not written into the federal patents. Rather, they were generally applicable to all similarly situated land within the state, without regard to whether the source of the title was the federal government, the state government, or some prior sovereign such as the King of Spain or the Republic of Mexico. The federal government disposed of its land largely without restriction, allowing market forces to determine the use of the resource. The contrast between this system of resource management and the present system of spectrum management is obvious.

2. *Disposition of water rights*

The disposition of water in the Western states is even more instructive. When the West was settled, the received common law also contained the rudiments of a property system in water, but the system was unsuitable for the conditions of that arid territory. Working from the customary practices of the settlers, both miners and farmers, the courts fashioned a new property system for water that, in contrast to the received system, facilitated investment and allowed for reallocation through transfers. This property institution—the appropriation system—gave each user the right to use the quantity of water he had diverted to beneficial use, with the right to receive such quantity of water before any subsequent user could obtain water. The time-tenure was perpetual, but transfers were permitted, except as later restricted by statute in a few states. Legislation later strengthened the appropriation system by providing clear evidence of ownership through a permit system and, more recently, by improving the quantification of the right from an amount diverted in flow per second to the amount consumed in volumetric measure. Roughly seventy-five years were consumed in refining the appropriation system, and details continue to be supplied by both courts and legislatures.

What is important for present purposes, however, is not the evolution of the appropriation system but the response of the federal government to its creation. To understand the position of the government and the decision it made, we must recount a little history. As a result of the war with Mexico, the federal government acquired a vast territory extending from the Rio Grande River to the Pacific Ocean encompassing what is now Arizona, California, New Mexico, and Utah. The Treaty of Guadalupe Hidalgo[82] gave the government fee ownership of this territory—land, water, minerals, and all, except for titles granted by previous sovereigns and confirmed by the treaty. Yet for nearly twenty years, during the period when the outlines of the appropriation system were being developed, no provision was made for disposition of these resources. Finally, in a series of statutes enacted in 1866,[83] 1870,[84] and 1877,[85] the government in effect quitclaimed its title to water appropriated under the appropriation system, both retroactively and prospectively. Thus the government relinquished its title to a very valuable resource to individuals and firms who acquired the resource under a system of law developed independently of and outside of lawmaking institutions of the federal government.[86]

3. *General characteristics*

The list of resources once owned by the federal government but transferred to private ownership could be greatly extended; it would include silver, gold, copper, a number of other precious

82 Treaty with Mexico, Feb. 2, 1848, 9 Stat. 922 (1848), T.S. No. 207.

83 Act of July 26, 1866, ch. 262, § 9, 14 Stat. 251, 253.

84 Act of July 9, 1870, ch. 235, § 17, 16 Stat. 217, 218.

85 Act of Mar. 3, 1877, ch. 107, § 1, 19 Stat. 377.

86 The Government also quitclaimed its title to gold, silver, and other valuable metals to the miners. *See* Act of July 26, 1866, ch. 262, 14 Stat. 251.

and semiprecious metals, oil and gas, and many other minerals. While the statutes governing the disposition of the various resources differ in important details, all the statutes are commonly characterized by a tenure of substantial duration and absence of restraints on the manner of exploiting the resource. Thus, a federal oil and gas lessee who discovers oil or gas on his leasehold gains a tenure for "so long as oil is produced." Moreover, a federal oil and gas lease does not tell the lessee that he must drill, or that if he does drill he must drill in a specified location or to a certain depth, and it does not tell the lessee whether to refine the oil into aviation gasoline or fuel oil or to manufacture it into petrochemical products.

In summary, traditional policies support the transfer of government-owned or -controlled natural resources from the public sector of the economy to the private sector. Once in the private sector, market forces will operate to allocate the use of the resource.

B. Transition to a Market System

There remains the question of what portion of the spectrum is available to be "property-ized" and turned over to market allocation. The present regulatory system could, of course, be abandoned, and all of the spectrum between 50 MHz and 1,000 MHz could be given the attributes of property suggested in chapter III. A more conservative position, affording an opportunity for evaluation of the success of the property system and of the reliability of market operations, would be to choose smaller portions of the spectrum for an experiment or a series of experiments. We believe that four portions of the spectrum are especially suitable for transfer to an experimental market-allocation system.

1. Use of alternate VHF- and intermediate UHF-television channels

Present FCC regulations provide for the assignment of alternate VHF-TV channels and for the assignment of every sixth UHF channel in the same geographical area. Other radio services could use the unassigned channels without interfering with the television broadcasting and without disturbing the present system of television allocation and licensing. Since these unassigned channels represent virgin spectrum, they could be formed into TAS packages and auctioned off either for time tenure in perpetuity or under long-term leases. The UHF portion of the experiment could be confined to the seven lowest UHF-TV channels. Modification of existing UHF voice-communication equipment for operation at these frequencies could be done at relatively low cost. Using these channels, a 24-MHz band of spectrum could be made available for experimentation in New York City, an 18-MHz band in Chicago, and a 36-MHz band in Los Angeles. In smaller urban areas even more spectrum could be used.

2. Clearing of voice and UHF-television channels

Some spectrum could be cleared in the existing voice-channel allocations of 150 to 170 MHz and 450 to 470 MHz and in UHF-TV channels 14 and 15. These frequencies would then be formed into TAS combinations and auctioned off to the highest bidders. The procedure for clearing the frequencies could be as follows:

The government would announce that all existing voice-channels would expire at a predetermined date; licenses terminating before that date would be renewed only for the time remaining until the termination date. The present voice-channel licenses are not exclusive because the FCC has granted as many licenses for a given channel as there have been qualified applicants, with the result that some channels in a single metropolitan area have as many as fifty users. Therefore, the refusal to renew licenses should cause very little frustration of reasonable expectations, especially when present users can enter the auction and obtain a TAS package.

Channels 14 and 15 would be cleared for auctioning by giving the existing thirty-eight licensees (as of July 1, 1968) a choice between either a higher UHF channel and compensation for the costs of modifying their equipment[87] or a credit of, say $100,000

[87] The cost of "moving" to a different channel depends upon, for example, whether the TV station has been built or is in the process of completion, and the extent of the required change of channels. These costs are estimated to be a minimum of $100,000 in Federal Communications Commission, Report of Land Mobile Frequency Relief Committee, Jan. 19, 1968, at 3.

on the bid price for a full 6-MHz channel, either 14 or 15, to be bid for at the same time voice channels are auctioned. Therefore, a TV station currently operating on channel 14 or 15 would have to bid more than the total of all bids by individual voice-channel users for the channel, less $100,000, to maintain its right to use that channel.[88] For a station electing to move to a higher frequency, the appropriate agency would adjust its current input restrictions (antenna height, antenna location, and transmitter power) to compensate for the reduced area of coverage caused by using the higher frequency; that is, an attempt would be made to maintain the existing value of its FCC license.

It is possible that the above changes in all of the voice-channel bands would result in dislocations serious enough to reduce the value of the experiment. The experiment could be modified by maintaining the current status of existing voice-channel bands, auctioning only the newly created voice channels in UHF-TV channels 14 and 15; this would, however, reduce the value of the experiment by providing less information on the effects of market arrangements in spectrum use. For example, current voice-channel users pay zero pecuniary cost for use of "party line" channels, although they do pay substantial nonpecuniary queuing costs. Before purchasing exclusive experimental rights, they would compare the pecuniary costs of purchasing TAS packages to the nonpecuniary costs of waiting.

3. *Vesting FM rights*

A third way to acquire spectrum for a market experiment is to vest the output rights of present FM stations. The FM band is a strong candidate for any experiment designed to involve competition between different uses for TAS rights because the proximity of the

88 The UHF station would have to bid more than the total of all bids by individual users (less $100,000) because television broadcasting currently uses a band 6-MHz wide, while most voice-channel users would use a band only 25- to 200-kHz wide. Individual users might form associations to represent them in the auction, thus making their joint bid higher than the sum of their separate bids. This would make it less likely that the UHF operator would outbid the voice-channel users to acquire the exclusive rights to the 6-MHz band.

An association of individual users could serve other purposes as well; for example, the association could lease TAS rights to individual users. A wealth-maximizing association would price according to the amount of use by metering

FM band to existing TV and voice-channel bands makes the FM band adaptable to such other uses at relatively low modification costs.

Unlike the other experiments suggested, this one requires neither auctions nor dislocations to current FCC licensees. The existing rights of FM stations, now specified in terms of inputs (antenna height, antenna location, and transmitter power) would be redefined in terms of field-strength limits (TAS packages) and vested in the present station owners. To redefine the rights the appropriate federal agency would draw the approximate field-strength contour implied by the current FCC license rights, set area and frequency boundaries, and establish field-strength limits. An attempt would be made to maintain the existing value of the station in terms of audience, coverage, and reception. The only change involves the nature of rights held—the fully transferable and divisible TAS rights would replace the FCC input specifications.

There should be no objection to this experiment as a "giveaway." Investments in transmitters, receivers, business organization, and good will give rise to reasonable expectations of continuity of the FCC license despite the fact that the law requires what has been pro forma license renewal. Moreover, we have noted that vesting of "squatter's rights" in users has been traditional government policy for water and gold and that, if the property interest has been appropriately defined, such vesting does not interfere with reallocation of the resource through exchanges.

4. *Packing FM stations*

Another experiment in the FM band would involve clearing spectrum in which TAS packages could be created and auctioned to the highest-bidding users—TV, voice-channel, or FM—in the manner earlier outlined for experiments 1 and 2. By "packing" or

---

the use by each subscriber; this would lead to more business calls and fewer personal calls than are presently made on voice-channel frequencies.

Alternatively, manufacturing firms might purchase TAS packages and lease portions to customers along with the sale or lease of communications equipment in much the same way as computer services are presently sold. Financial companies could also purchase TAS rights and then lease them to users of communications systems.

squeezing existing FM stations into the 94- to 108-MHz band, as much as 6 MHz could be cleared for experimental purposes. The only modification of existing rights would be the specific frequencies used by the stations and, as before, an attempt would be made to maintain the existing value of current licensees' rights by adjusting other inputs such as power and antenna height. Also, licensees could be compensated out of the proceeds of the auction for their costs of changing frequencies, estimated at a minimum of $20,000 per station.

Unfortunately, this experiment cannot be performed in every major region of the nation. The FCC's current practice is to grant FM licenses for alternate 200-kHz channels, with 200-kHz guardbands. Even the highest quality FM receivers currently manufactured require these 200-kHz guardbands.[89] In the case of the Los Angeles region, packing present FM stations into the 94- to 108-MHz band would reduce guardbands below the 200-kHz constraint, thus reducing the quality of FM reception. Consequently, the packing experiment could not be usefully conducted in the Los Angeles region.

Generally, this experiment may be feasible only in less densely populated areas where the value of spectrum undoubtedly is lower. This does not mean the experiment should not be considered in a few heavily populated regions. In some cases possibly less than 6 MHz (the amount presently used for television broadcasting) could be cleared, thus allowing at least for competition among non-TV services. These new TAS rights might also be created in cities where significant population growth is anticipated.[90]

89 The reason is simply that equipment manufacturers do not now have any incentives to economize on the use of the scarce spectrum by producing higher-quality receivers.

90 The two FM experiments need not be considered in all-or-nothing terms. A city-by-city approach is appropriate. For cities where relatively little spectrum can be cleared, the vesting experiment might be preferable. Or the two experiments could be combined: Existing stations could be packed into a smaller total band, with their input rights being converted into output terms, thus supplementing the TAS packages that would be formed from the cleared portion of the FM band.

CONCLUSION

The present management of the electromagnetic spectrum has aroused sufficient concern to justify experiments in turning a portion of the spectrum over to the market for allocation. The success of such experiments depends on the creation of a property system in the spectrum that makes exchanges possible without undue transaction and enforcement costs. We have articulated a property system that we believe satisfies these criteria, and we have demonstrated that traditional policy supports the release of government-controlled resources to the private sector. Lastly, we have indicated portions of the spectrum that could be immediately transferred to the market system with relatively low dislocation costs. Our immediate purpose is to induce a change in the allocation of a particular resource; but we hope that this essay will encourage engineers, economists, and lawyers concerned with resource development and allocation to reflect critically on the property rights assigned to other resources, to the end of restructuring those rights where such changes would promote economic efficiency through greater reliance on market processes.

APPENDIX

Figure A-1 shows that TAS owners have incentives to keep their antennas sufficiently low that the radio horizons fall within their own TAS areas. Figure A-1 shows TAS owner A's signal field strength as a function of distance from his antenna for four combinations of antenna height and transmitter power that just meet the X v/m constraint. The corner in each curve occurs at the radio horizon for that antenna; it represents the transition from the free-space attenuation rate to the much more rapid shadow-zone attenuation rate. The higher curves correspond to *lower* antennas with transmitter power *increased* to compensate for the greater amount of shadow-zone attenuation so that the field strength at the boundary is X v/m in each case.

Consider first the solid portion of the lowest curve, labeled I; this corresponds to the highest antenna for which the horizon lies inside the *AB* boundary. Any higher antenna with the same transmitter power will extend *A*'s radio horizon into *B*'s area, and will increase the field strength in area *B*, as indicated by the dashed line; *B*'s service area will then represent a smaller percentage of

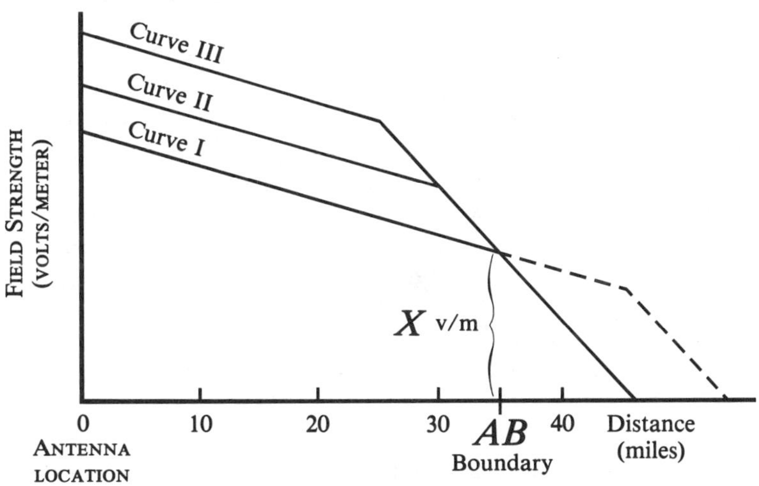

FIGURE A-1. Field strength as a function of distance from transmitting antenna for combinations of antenna height and transmitter power that produce field strength X v/m at a distance of 35 miles from the transmitting antenna

TABLE A-1. DATA UNDERLYING FIGURE A-1

Parameters *(Curves)*	Transmitting Antenna Height *(Feet)*	Transmitting Power *(Kilowatts)*	Service Area Radius *(Miles)*	Percentage—Service Area
I—solid line	497.0	1.0	22.4	41
I—dashed line ...	868.0	1.0	22.4	41
II	312.5	5.0	28.0	64
III	112.5	58.3	28.0	64

NOTE: Empirical data in P. RICE, A. LONGLEY, K. NORTON & A. BARSIS, TRANSMISSION LOSS PREDICTIONS FOR TROPOSPHERIC COMMUNICATIONS CIRCUITS, U.S. NATIONAL BUREAU OF STANDARDS, TECHNICAL NOTE 101, at 6 (rev. 1966), were used to obtain the results in this table. However, because field strength is very sensitive to the nature of the intervening terrain, the performance of any single radio circuit may deviate markedly from these averaged results. The calculations for all four curves assume a frequency of 400 MHz, a receiving antenna height of 6 feet, a signal-to-interference ratio at the edge of the service area of 1,000:1, and an interfering signal that attenuates at the shadow-zone rate.

his specified TAS area dimension than indicated in table 3. *A*'s field strength within his area, however, is not changed by raising his antenna because his entire area was already within the horizon. Therefore, increasing the height of *A*'s antenna so that its horizon is beyond the *AB* boundary cannot improve *A*'s service area within his TAS area.

Now compare curve I with curve II, which represents a lower antenna with a higher transmitter power. Because of the more powerful transmitter, the resulting field strength will be higher at every point within *A*'s TAS area, and because of the closer radio horizon, the *X* v/m limit at the *AB* boundary will not be exceeded; consequently, *A* obtains a larger service area without diminishing *B*'s service area. Thus a TAS owner achieves larger service areas with antennas whose horizons are within, rather than beyond, his TAS area; the benefit to neighboring TAS owners is also a benefit to *A*. Curve III represents a still lower antenna and higher transmitter power. The fraction of *A*'s TAS area that is service area will be no larger (see table A-1), but the higher field strengths in the center of the area may result in higher-quality communications there. In general, the exact combination of antenna height, transmitter power, and resulting service area that a TAS owner adopts will depend upon the value of servicing the area compared with the costs of the necessary antenna and transmitter.

RECOMMENDED READING

Baker, Joan. "Free Speech and Federal Control: The U.S. Approach to Broadcasting Regulation." *Modern Law Review* 39 (1976):147–61.

Barrow, Roscoe L. "Fairness Doctrine: A Double Standard for Electronic and Print Media." *Hastings Law Journal* 26 (1975):659–708.

Bazelon, David. "FCC Regulation of the Telecommunications Press." *Duke Law Journal* 1975:213–51.

Boros, J. S. "Communications Law—License Renewal Challenges. Regulation of Programming." *Annual Survey of American Law* 1973/74 (1974):597–264.

Bowdre, Karen O. "FCC v. National Citizens Committee for Broadcasting—A Closer Step to Complete Media Regulation." *Cumberland Law Review* 9 (1979):851–68.

Brenner, Daniel L. "Limits of Broadcast Self-Regulation Under the First Amendment." *Federal Communications Bar Journal* 28 (1975):1–62.

Brenner, D. "Censoring the Airwaves." *Barrister* 5 (1978):11.

Brown, Nicholas K. "Subscription Television Controversy: A Continuing Symptom of Federal Communication Commission Ills." *Federal Communications Bar Journal* 24 (1970–71): 259–82.

Cass, Ronald A. "Regulation of Pay-Cable and Closed Circuit Movies: No Room in the Wasteland." *University of Chicago Law Review* 40 (1973):600–16.

Cheek, Leslie. "Analysis of Proposals to Deregulate Commercial Radio Broadcasting." *Federal Communications Bar Journal* 25 (1972):1–52.

Clark, David G., and Hutchison, E. R. "Self-Censorship in Broadcasting—The Cowardly Lions." *New York Law Forum* 18 (1972):18–31.

Dugan, Thomas E. "FCC—Constitutional Right to Free Speech—Limp Libidinal Language." *Marquette Law Review* 61 (1978):534–43.

Everett, J. Diane. "Broadcast Media Regulation: Death Knell of the Fairness Doctrine." *University of Florida Law Review* 27 (1975):607–13.

Fastow, J. N. "Competition, Competitors and the Government's Suit Against the Television Networks." *Antitrust Bulletin* 22 (1977):517–38.

Fine, J. David. "Whither Goes the Wire? The Extent of Federal Competence to Regulate CATV." *McGill Law Journal* 18 (1972):615–23.

Fournet, Sandra M. "Pay Television: The Pendulum Swings Toward Deregulation." *Washburn Law Journal* 18 (1978):86–104.

Freeman, Harrop, and Edelstein, Stewart. "Political Campaigning and the Airways." *Pepperdine Law Review* 2 (1974): 178–208.

Goldberg, H., and Couzens, M. "'Peculiar Characteristics': An Analysis of the First Amendment Implications of Broadcast Regulation." *Federal Communications Law Journal* 31 (1978):1–50.

Hagelin, Theodore M. "First Amendment Stake in New Technology: The Broadcast Cable Controversy." *University of Cincinnati Law Review* 44 (1975):427–524.

Howard, Herbert H. "Multiple Broadcast Ownership: Regulatory History." *Federal Communications Bar Journal* 27 (1974):1–70.

Loevinger, Lee. "Is There Intelligent Life in Washington?" *Federal Communications Bar Journal* 29 (1976):189–205.

Marks, Richard D. "Broadcasting and Censorship: First Amendment Theory After Red Lion." *George Washington Law Review* 38 (1970):974–1005.

Minasian, Jora R. "Property Rights in Radiation: An Alternative Approach to Radio Frequency Allocation." *Journal of Law and Economics* 18 (1975):221–72.

Mustin, Scott H. "FCC Content Regulation of Cable-Pay Television: The Threat of Pacifica." *Cumberland Law Review* 9 (1979):811–29.

Pember, Don R. "Broadcaster and the Public Interest: A Proposal to Replace an Unfaithful Servant." *Loyola University Law Review (LA)* 4 (1971):83–111.

Price, Monroe E. "Requiem for the Wired Nation: Cable Rule-Making at the FCC." *Virginia Law Review* 61 (1975):541–77.

Robinson, Glen O. "Federal Communications Commission: An Essay on Regulatory Watchdogs." *Virginia Law Review* 64 (1978):169–262.

Schenkkan, P. M. "Power in the Marketplace of Ideas: The Fairness Doctrine and the First Amendment." *Texas Law Review* 52 (1974):727–72.

Smulyan, Jefferey H. "Power to Some People: The FCC's Clear Channel Allocation Policy." *Southern California Law Review* 44 (1971):811–847.

Stull, Dr. Mark A., and Alexander, George. "Passive Use of the Radio Spectrum for Scientific Purposes and the Frequency Allocation Process." *Journal of Air Law* 43 (1977):459–534.

Swartz, John L. "Fairness for Whom? Administration of the Fairness Doctrine 1969–70." *Boston College Industrial & Commercial Law Review* 14 (1973):457–70.

Sweeney, Walter H. "Regulation of Television Program Content by the Federal Communications Commission." *University of Richmond Law Review* 8 (1974):233–44.

Volner, I. D. "Broadcast Regulation: Is There Too Much 'Public' in the 'Public Interest'?" *University of Cincinnati Law Review* 43 (1974):267–89.

Werner, Robert L. "A Lawyer Looks at Our Communications Policy." *Jurimetrics* 11 (1970):81–86.

Weyant, Piers J. "Administrative Law—The FCC's Fairness Doctrine and the First Amendment." *Journal of Public Law* 19 (1970):129–37.

Wilson, J. L. "Fairness Doctrine: Big Brother in the Newsroom." *American Bar Association Journal* 61 (1975):1492–94.

Wilson, R. A. "Antitrust, Economics, and the Media: The Next Voice You Hear May Be Your Own!" *Antitrust Law and Economic Review* 10 (1978):35–51.

ABOUT THE AUTHORS

Arthur S. De Vany, A.B. 1963, M.A. 1965, University of California, Los Angeles. Research Economist, General Electric Company, TEMPO Center for Advanced Studies.

Ross D. Eckert, A.B. 1963, M.A. 1964, Ph.D. 1968, University of California, Los Angeles. Research Economist, General Electric Company, TEMPO Center for Advanced Studies.

Charles J. Meyers, B.A. 1949, Rice University; LL.B. 1949, University of Texas; LL.M. 1953, J.S.D. 1964, Columbia University. Professor of Law, Stanford University.

Donald J. O'Hara, B.S. 1964, California Institute of Technology; M.A. 1966, University of California, Los Angeles. Research Economist, General Electric Company, TEMPO Center for Advanced Studies.

Richard C. Scott, B.S. 1951, University of Michigan. Research Engineer, General Electric Company, TEMPO Center for Advanced Studies.

The Cato Papers

1.	*Left and Right: The Prospects for Liberty,* by Murray N. Rothbard	$2.00
2.	*The Fallacy of the Mixed Economy: An "Austrian" Critique of Conventional Economics and Government Policy,* by Stephen C. Littlechild	4.00
3.	*Unemployment and Monetary Policy: Government as Generator of the "Business Cycle,"* by Friedrich A. Hayek	2.00
4.	*Individualism and the Philosophy of the Social Sciences,* by Murray N. Rothbard	2.00
5.	*National Income Statistics: A Critique of Macroeconomic Aggregation,* by Oskar Morgenstern	2.00
6.	*A Tiger by the Tail: The Keynesian Legacy of Inflation,* by Friedrich A. Hayek	4.00
7.	*Strategic Disengagement and World Peace: Toward a Noninterventionist American Foreign Policy,* by Earl C. Ravenal	2.00
8.	*A Theory of Strict Liability: Toward a Reformulation of Tort Law,* by Richard A. Epstein	4.00
9.	*Is Government the Source of Monopoly? Market Structure and the Competitive Process,* by Yale Brozen	2.00